# 2016 SQA Past Papers & Hodder Gibson Model Papers With Answers

## Advanced Higher

# CHEMISTRY

2015 Specimen Question Paper,
Model Papers & 2016 Exam



SQA Advanced Higher CHEMISTRY

This book contains the official 2015 SQA Specimen Question Paper and the 2016 Exam for Advanced Higher Chemistry, with associated SQA-approved answers modified from the official marking instructions that accompany the paper.

In addition the book contains model papers, together with answers, plus study skills advice. These papers, some of which may include a limited number of previously published SQA questions, have been specially commissioned by Hodder Gibson, and have been written by experienced senior teachers and examiners in line with the new Advanced Higher for CfE syllabus and assessment outlines. This is not SQA material but has been devised to provide further practice for Advanced Higher examinations in 2016 and beyond.




Hodder Gibson is grateful to the copyright holders, as credited on the final page of the Answer Section, for permission to use their material. Every effort has been made to trace the copyright holders and to obtain their permission for the use of copyright material. Hodder Gibson will be happy to receive information allowing us to rectify any error or omission in future editions.

Hachette UK's policy is to use papers that are natural, renewable and recyclable products and made from wood grown in sustainable forests. The logging and manufacturing processes are expected to conform to the environmental regulations of the country of origin.

Orders: please contact Bookpoint Ltd, 130 Park Drive, Milton Park, Abingdon, Oxon OX14 4SE. Telephone: (44) 01235 827720. Fax: (44) 01235 400454. Lines are open 9.00–5.00, Monday to Saturday, with a 24-hour message answering service. Visit our website at www.hoddereducation.co.uk. Hodder Gibson can be contacted direct on: Tel: 0141 333 4650; Fax: 0141 404 8188; email: hoddergibson@hodder.co.uk

This collection first published in 2016 by
Hodder Gibson, an imprint of Hodder Education,
An Hachette UK Company
211 St Vincent Street
Glasgow G2 5QY





Typeset by Aptara, Inc.

Printed in the UK

A catalogue record for this title is available from the British Library

ISBN: 978-1-4718-9076-5

3 2 1

2017 2016

# Introduction

## Study Skills – what you need to know to pass exams!

### Pause for thought

Many students might skip quickly through a page like this. After all, we all know how to revise. Do you really though?

*Think about this:*

"IF YOU ALWAYS DO WHAT YOU ALWAYS DO, YOU WILL ALWAYS GET WHAT YOU HAVE ALWAYS GOT."

Do you like the grades you get? Do you want to do better? If you get full marks in your assessment, then that's great! Change nothing! This section is just to help you get that little bit better than you already are.

There are two main parts to the advice on offer here. The first part highlights fairly obvious things but which are also very important. The second part makes suggestions about revision that you might not have thought about but which WILL help you.

### Part 1

DOH! It's so obvious but ...

*Start revising in good time*

Don't leave it until the last minute – this will make you panic.

Make a revision timetable that sets out work time AND play time.

*Sleep and eat!*

Obvious really, and very helpful. Avoid arguments or stressful things too – even games that wind you up. You need to be fit, awake and focused!

*Know your place!*

Make sure you know exactly **WHEN and WHERE** your exams are.

*Know your enemy!*

**Make sure you know what to expect in the exam.**

How is the paper structured?

How much time is there for each question?

What types of question are involved?

Which topics seem to come up time and time again?

Which topics are your strongest and which are your weakest?

Are all topics compulsory or are there choices?

*Learn by DOING!*

There is no substitute for past papers and practice papers – they are simply essential! Tackling this collection of papers and answers is exactly the right thing to be doing as your exams approach.

### Part 2

People learn in different ways. Some like low light, some bright. Some like early morning, some like evening / night. Some prefer warm, some prefer cold. But everyone uses their BRAIN and the brain works when it is active. Passive learning – sitting gazing at notes – is the most INEFFICIENT way to learn anything. Below you will find tips and ideas for making your revision more effective and maybe even more enjoyable. What follows gets your brain active, and active learning works!

*Activity 1 – Stop and review*

Step 1

When you have done no more than 5 minutes of revision reading STOP!

Step 2

Write a heading in your own words which sums up the topic you have been revising.

Step 3

Write a summary of what you have revised in no more than two sentences. Don't fool yourself by saying, "I know it, but I cannot put it into words". That just means you don't know it well enough. If you cannot write your summary, revise that section again, knowing that you must write a summary at the end of it. Many of you will have notebooks full of blue/black ink writing. Many of the pages will not be especially attractive or memorable so try to liven them up a bit with colour as you are reviewing and rewriting. **This is a great memory aid, and memory is the most important thing.**

### *Activity 2 – Use technology!*

Why should everything be written down? Have you thought about "mental" maps, diagrams, cartoons and colour to help you learn? And rather than write down notes, why not record your revision material?

What about having a text message revision session with friends? Keep in touch with them to find out how and what they are revising and share ideas and questions.

Why not make a video diary where you tell the camera what you are doing, what you think you have learned and what you still have to do? No one has to see or hear it, but the process of having to organise your thoughts in a formal way to explain something is a very important learning practice.

Be sure to make use of electronic files. You could begin to summarise your class notes. Your typing might be slow, but it will get faster and the typed notes will be easier to read than the scribbles in your class notes. Try to add different fonts and colours to make your work stand out. You can easily Google relevant pictures, cartoons and diagrams which you can copy and paste to make your work more attractive and **MEMORABLE**.

### *Activity 3 – This is it. Do this and you will know lots!*

#### Step 1

In this task you must be very honest with yourself! Find the SQA syllabus for your subject (www.sqa.org.uk). Look at how it is broken down into main topics called MANDATORY knowledge. That means stuff you MUST know.

#### Step 2

BEFORE you do ANY revision on this topic, write a list of everything that you already know about the subject. It might be quite a long list but you only need to write it once. It shows you all the information that is already in your long-term memory so you know what parts you do not need to revise!

#### Step 3

Pick a chapter or section from your book or revision notes. Choose a fairly large section or a whole chapter to get the most out of this activity.

With a buddy, use Skype, Facetime, Twitter or any other communication you have, to play the game "If this is the answer, what is the question?". For example, if you are revising Geography and the answer you provide is "meander", your buddy would have to make up a question like "What is the word that describes a feature of a river where it flows slowly and bends often from side to side?".

Make up 10 "answers" based on the content of the chapter or section you are using. Give this to your buddy to solve while you solve theirs.

#### Step 4

Construct a wordsearch of at least 10 × 10 squares. You can make it as big as you like but keep it realistic. Work together with a group of friends. Many apps allow you to make wordsearch puzzles online. The words and phrases can go in any direction and phrases can be split. Your puzzle must only contain facts linked to the topic you are revising. Your task is to find 10 bits of information to hide in your puzzle, but you must not repeat information that you used in Step 3. DO NOT show where the words are. Fill up empty squares with random letters. Remember to keep a note of where your answers are hidden but do not show your friends. When you have a complete puzzle, exchange it with a friend to solve each other's puzzle.

#### Step 5

Now make up 10 questions (not "answers" this time) based on the same chapter used in the previous two tasks. Again, you must find NEW information that you have not yet used. Now it's getting hard to find that new information! Again, give your questions to a friend to answer.

#### Step 6

As you have been doing the puzzles, your brain has been actively searching for new information. Now write a NEW LIST that contains only the new information you have discovered when doing the puzzles. Your new list is the one to look at repeatedly for short bursts over the next few days. Try to remember more and more of it without looking at it. After a few days, you should be able to add words from your second list to your first list as you increase the information in your long-term memory.

## FINALLY! Be inspired...

Make a list of different revision ideas and beside each one write **THINGS I HAVE** tried, **THINGS I WILL** try and **THINGS I MIGHT** try. Don't be scared of trying something new.

And remember – "FAIL TO PREPARE AND PREPARE TO FAIL!"

# Advanced Higher Chemistry

## The course

The main aims of the Advanced Higher Chemistry course are for learners to:

- develop a critical understanding of the role of chemistry in scientific issues and relevant applications, including the impact these could make on the environment/society
- extend and apply knowledge, understanding and skills of chemistry
- develop and apply the skills to carry out complex practical scientific activities, including the use of risk assessments, technology, equipment and materials
- develop and apply scientific inquiry and investigative skills, including planning and experimental design
- develop and apply analytical thinking skills, including critical evaluation of experimental procedures in a chemistry context
- extend and apply problem-solving skills in a chemistry context
- further develop an understanding of scientific literacy, using a wide range of resources, in order to communicate complex ideas and issues and to make scientifically informed choices
- extend and apply skills of independent/autonomous working in chemistry.

### *How the course is assessed*

To gain the course award:

- You must pass three units which are assessed internally on a pass/fail basis. The units are:
  1. Inorganic and Physical Chemistry
  2. Organic Chemistry and Instrumental Analysis
  3. Researching Chemistry
- You must submit a project which is marked externally by the SQA and is worth a total of 30 marks.
- You must sit the Advanced Higher Chemistry exam paper which is set and marked by SQA and is worth 100 marks.

The course award is graded A–D, the grade being determined by the total mark you score in the examination and the mark you gain in the project.

### *The examination*

The examination consists of one paper which has two sections and lasts 2 hours 30 minutes. Section 1 is multiple choice and is worth 30 marks. Section 2 comprises extended answer questions and is worth 70 marks.

Further details can be found in the Advanced Higher Chemistry section on the SQA website: http://www.sqa.org.uk/sqa/48459.html

## Key tips for your success

### *Practise! Practise! Practise!*

The key to exam success in Chemistry is to prepare by answering questions regularly. Use the questions as a prompt for further study. If you find that you cannot answer a question, review your notes and/or textbook to help you find the necessary knowledge to answer the question. This is a very quick way of finding out what you can or cannot do and is a better use of your time than copying notes passively, which is a common trap many students fall into!

### *The data booklet*

The data booklet contains formulae and useful data which you will have to use in the exam. Although you might think that you have a good memory for chemical data (such as the symbols or atomic masses of elements), it is always worthwhile to check using the data booklet.

### *Calculations*

In preparation for the exam, ensure that you recognise the different calculation types and know when to use the formulae from the data booklet:

- Energy, frequency and wavelength

  $E = Lhf$

  $c = f\lambda$

- Using the pH equations

  strong acids: $pH = -\log[H^+]$

  weak acids: $pH = \frac{1}{2}pK_a - \frac{1}{2}\log c$

  buffers: $pH = pK_a - \log\frac{[acid]}{[salt]}$

- Thermodynamics

  $\Delta S = \Sigma S^0 \text{products} - \Sigma S^0 \text{reactants}$

  $\Delta H = \Sigma H^0 \text{products} - \Sigma H^0 \text{reactants}$

  $\Delta G^0 = \Delta H^0 - T\Delta S^0$

Remember that you have to make the units for $\Delta S$ compatible with $\Delta H$ and $\Delta G$. This can be done by dividing $\Delta S$ by 1000 so that it has the units $kJ\,K^{-1}\,mol^{-1}$ **or** by multiplying both $\Delta H$ and $\Delta G$ by 1000 so that they have the units $J\,mol^{-1}$.

- Kinetics: calculating the rate constant, k, from rate equations
- Synthesis: calculating the % yield
- Solutions: calculating % solution by mass, % solution by volume and ppm
- Calculations from balanced chemical equations (stoichiometry)
- Calculating the empirical formula

You will encounter these calculations in the exam so it is worth spending time practising them as part of your revision to ensure that you are familiar with the routines for solving these problems. Even if you are not sure how to attempt a calculation question, show your working! You will be given credit for calculations which are relevant to the problem being solved.

### *"Explain" questions*

You will encounter questions which ask you to *explain your answer*. Take your time when answering such questions and try to give a clear explanation. If possible, use diagrams or chemical equations; they can really bring an answer to life.

### *Applying your knowledge of practical chemistry*

As part of your Advanced Higher Chemistry experience, you should have had plenty of practice carrying out experiments using standard lab equipment. You will also have had opportunities to evaluate your results from experiments. In the Advanced Higher exam, you are expected to be familiar with the techniques and apparatus given in the following table. This table has been taken from the SQA AH Chemistry Course Support Notes, April 2015 (http://www.sqa.org.uk/files_ccc/AHCUSNChemistry.pdf).

| *Learners would benefit from being familiar with the following apparatus, practical techniques and activities:* | *Learners should be able to process experimental results by:* |
| --- | --- |
| • digital balance<br>• Buchner or Hirsch or sintered glass funnel<br>• glassware with ground glass joints ('Quickfit' or similar)<br>• thin-layer chromatography apparatus<br>• weighing by difference and gravimetric analysis<br>• preparing a standard solution<br>• using a reference or control or blank determination<br>• carrying out a complexometric titration<br>• carrying out a back titration<br>• using a colorimeter or visible spectrophotometer and carrying out dilution to prepare a calibration graph<br>• distilling<br>• refluxing<br>• using vacuum filtration methods<br>• recrystallising<br>• determining % yield experimentally<br>• using thin-layer chromatography<br>• using melting point apparatus and mixed melting point determination<br>• using a separating funnel and solvent extraction | • representing experimental data using a scatter graph<br>• sketching lines or curves of best fit<br>• calculating mean values for experiments<br>• identifying and eliminating rogue points from the analysis of results<br>• qualitative appreciation of the relative accuracy of apparatus used to measure the volume of liquids (Learners would be expected to know that the volume markings on beakers provide only a rough indication of volume. While measuring cylinders generally provide sufficient accuracy for preparative work, for analytic work, burettes, pipettes and volumetric flasks are more appropriate.)<br>• appreciating that when a measurement has been repeated, any variations in the value obtained give an indication of the reproducibility of the technique<br>• knowing that the uncertainty associated with a measurement can be indicated in the form, *measurement ± uncertainty* (Learners are not expected to conduct any form of quantitative error analysis.)<br>• quantitative stoichiometric calculations<br>• interpretation of spectral data |

### *Analysis of data*

From your experience of working with experimental data in both Higher and Advanced Higher Chemistry, you should know how to calculate averages, eliminate rogue data and how to draw and interpret graphs (scatter and best fit line/curve).

It is common in Chemistry exams to be presented with titration data such as shown in the following table.

| *Titration* | *Volume of solution/cm$^3$* |
|---|---|
| 1 | 26·0 |
| 2 | 24·1 |
| 3 | 39·0 |
| 4 | 24·2 |
| 5 | 24·8 |

Result 1 is a rough titration which is not accurate.

Results 2 and 4 could be used to calculate an average volume (= 24·15 cm$^3$).

Result 3 is a rogue result and should be ignored.

Result 5 cannot be used to calculate the average volume as it is too far from 24·1 and 24·2. In other words, it is not accurate.

### *Open-ended questions*

Real-life chemistry problems rarely have a fixed answer. In the Advanced Higher exam you will encounter two three-mark questions that are open-ended; this means that there is more than one "correct" answer. You will recognise these from the phrase *using your knowledge of chemistry* in the question. *To tackle these questions, focus on the chemistry*. For example, if you are shown a molecule, look out for familiar functional groups. Perhaps you could answer the question by discussing the typical reactions and properties of these functional groups. If it is an organic molecule, you could discuss spectroscopic techniques and the likely results you would get if you were to analyse the molecule by NMR or IR. Other questions will require more specific knowledge of the AH course. Treat these as an opportunity to show the examiner how much you know about the topic of interest. However, ensure your answer actually answers the question! It is also worth noting that you will only gain credit for answers that are at Advanced Higher level. Answers at Higher or National 5 will be ignored and will gain zero marks.

## Good luck!

If you have followed the advice given in this introduction you will be well prepared for the Advanced Higher exam. When you sit the exam, take your time and use the experience as an opportunity to show the examiner how much you know. And good luck!

ADVANCED HIGHER

# 2015 Specimen Question Paper

**SQ05/AH/02**

# Chemistry
# Section 1 — Questions

Date — Not applicable

Duration — 2 hours 30 minutes

Instructions for the completion of Section 1 are given on *Page two* of your question and answer booklet SQ05/AH/01.

Record your answers on the answer grid on *Page three* of your question and answer booklet.

Reference may be made to the Chemistry Higher and Advanced Higher Data Booklet.

Before leaving the examination room you must give your question and answer booklet to the Invigilator; if you do not you may lose all the marks for this paper.

## SECTION 1 — 30 marks

**Attempt ALL questions**

**1.** An element X forms an ion, $X^{3+}$, which contains 55 electrons.

In which block of the Periodic Table would element X be found?

A s

B p

C d

D f

**2.** Which one of the following metal salts will emit radiation of the highest frequency when placed in a Bunsen flame?

A Copper(II) chloride

B Potassium chloride

C Barium chloride

D Lithium chloride

**3.** The following diagram represents a square-planar structure.

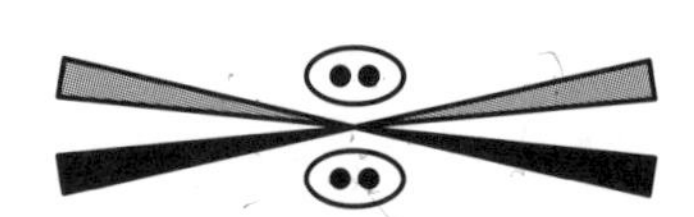

Where (black wedge) and (grey wedge) represent bonding electron pairs and (●●) represents a non-bonding electron pair (lone pair).

Which of the following species could have the structure shown above?

A $SF_4$

B $NH_4^+$

C $XeF_4$

D $AlH_4^-$

**4.** A complex ion with the name hexaamminenickel(II) will have the formula

A $[Ni(NH_2)_6]^{2+}$

B $[Ni(NH_3)_6]^{2+}$

C $[Ni(NH_3)_6]^{4-}$

D $[Ni(NH_4)_6]^{2+}$.

**5.** The reaction

$$CO(g) + 3H_2(g) \rightleftharpoons CH_4(g) + H_2O(g)$$

has an equilibrium constant of 3·9 at 950 °C.

The equilibrium concentrations of CO(g), $H_2(g)$ and $H_2O(g)$ at 950 °C are given in the table.

| *Substance* | *Equilibrium concentration/mol $l^{-1}$* |
|---|---|
| CO(g) | $5{\cdot}0 \times 10^{-2}$ |
| $H_2(g)$ | $1{\cdot}0 \times 10^{-2}$ |
| $H_2O(g)$ | $4{\cdot}0 \times 10^{-3}$ |

What is the equilibrium concentration of $CH_4(g)$, in mol $l^{-1}$, at 950 °C?

A $4{\cdot}9 \times 10^{-1}$

B $3{\cdot}1 \times 10^{-5}$

C $4{\cdot}9 \times 10^{-5}$

D $2{\cdot}0 \times 10^{-7}$

**6.** Which of the following decreases when an aqueous solution of ethanoic acid is diluted?

A pH

B $[H^+]$

C $pK_a$

D $K_a$

**7.** The pH of a buffer solution prepared by mixing equal volumes of 0·1 mol $l^{-1}$ ethanoic acid and 0·2 mol $l^{-1}$ sodium ethanoate is

A 2·1

B 2·7

C 4·5

D 5·1.

**8.** The graph below shows the pH changes when 0·1 mol $l^{-1}$ ammonia solution is added to 50 $cm^3$ of 0·1 mol $l^{-1}$ hydrochloric acid solution.

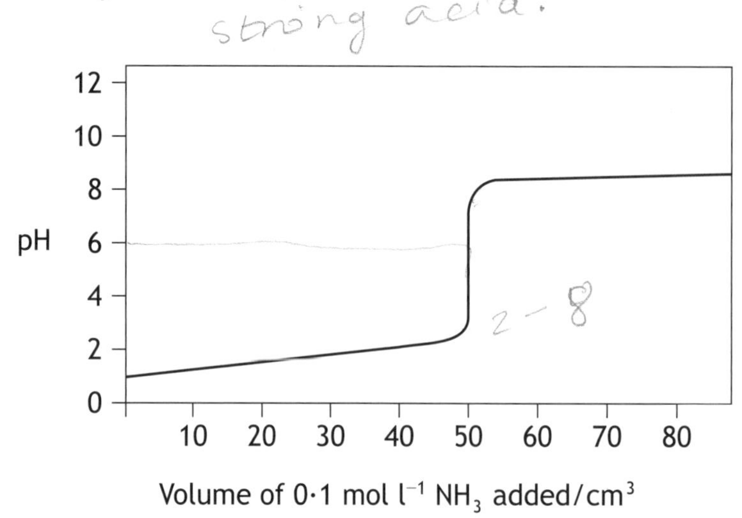


Which line in the table shows an indicator which is **not** suitable for use in determining the equivalence point for the above reaction?

| | *Indicator* | *pH range of indicator* |
|---|---|---|
| A | methyl orange | 3·1 — 4·4 |
| B | bromophenol red | 5·2 — 6·8 |
| C | bromothymol blue | 6·0 — 7·6 |
| D | phenolphthalein | 8·3 — 10·0 |

**9.** Ethanoic acid is a weak acid and hydrochloric acid is a strong acid. Which of the following is **not** true?

A The pH of 0·1 mol $l^{-1}$ hydrochloric acid is 1.

B 20·0 $cm^3$ of 0·1 mol $l^{-1}$ sodium hydroxide is exactly neutralised by 20·0 $cm^3$ of 0·1 mol $l^{-1}$ ethanoic acid.

C The pH of 0·1 mol $l^{-1}$ hydrochloric acid is lower than that of 0·1 mol $l^{-1}$ ethanoic acid.

D The $K_a$ value for ethanoic acid is greater than that of hydrochloric acid.

**10.** The standard enthalpy of formation of strontium chloride is the enthalpy change for which of the following reactions?

A $Sr(g) + Cl_2(g) \rightarrow SrCl_2(s)$

B $Sr(s) + Cl_2(g) \rightarrow SrCl_2(s)$

C $Sr^{2+}(g) + 2Cl^-(g) \rightarrow SrCl_2(s)$

D $Sr^{2+}(aq) + 2Cl^-(aq) \rightarrow SrCl_2(s)$

**11.** Which of the following alcohols would have the greatest entropy at 90 °C?

A Propan-1-ol

B Butan-1-ol

C Propan-2-ol

D Butan-2-ol

**12.** For any liquid, $\Delta S_{vaporisation} = \dfrac{\Delta H_{vaporisation}}{T_b}$

where $T_b$ = boiling point of that liquid.

For many liquids, $\Delta S_{vaporisation} = 88\ J\ K^{-1}\ mol^{-1}$.

Assuming that this value is true for water and that its $\Delta H_{vaporisation} = 40{\cdot}6\ kJ\ mol^{-1}$, then the boiling point of water is calculated as

A 0·46 K

B 2·17 K

C 373 K

D 461 K.

**13.** The order of a reactant in a reaction

A can only be obtained by experiment

B determines the speed of the overall reaction

C is determined by the stoichiometry involved

D is the sequence of steps in the reaction mechanism.

**14.** A suggested mechanism for the reaction

$2X + Y \rightarrow X_2Y$

is a two-step process

$X + Y \rightarrow XY$ (slow)
$XY + X \rightarrow X_2Y$ (fast)

This mechanism is consistent with which of the following rate equations?

A Rate = k[XY]

B Rate = k[X][Y]

C Rate = $k[X]^2[Y]$

D Rate = k[X][XY]

**15.** What volume of $0{\cdot}2\,mol\,l^{-1}$ potassium sulfate is required to make, by dilution with water, one litre of a solution with a **potassium** ion concentration of $0{\cdot}1\,mol\,l^{-1}$?

A $500\,cm^3$

B $400\,cm^3$

C $250\,cm^3$

D $100\,cm^3$

**16.** The end-on overlap of two atomic orbitals lying along the axis of a bond leads to

A hydridisation

B a sigma bond

C a pi bond

D a double bond.

**17.** Which of the following has nucleophilic properties?

A Na

B $Br^+$

C $CH_3^+$

D $NH_3$

**18.** Carbonyl groups in aldehydes and ketones react with HCN and the product can then be hydrolysed forming a 2-hydroxycarboxylic acid as shown in the equation below.

$$H_2C{=}O \xrightarrow{HCN} H{-}CH(OH){-}C{\equiv}N \xrightarrow{H^+(aq)} H{-}CH(OH){-}COOH$$

When the final product is 2-hydroxy-2-methylbutanoic acid, the starting carbonyl compound is

A propanol

B propanone

C butanal

D butanone.

**19.** Which of the following is a tertiary haloalkane?

A $CHCl_3$

B $(CH_3)_3CCl$

C $(CH_2Cl)_3CH$

D $(CH_3)_3CCH_2Cl$

**20.** The structures of three alcohols, P, Q and R are shown.

```
      H   H   OH
      |   |   |
P  H—C—C—C—H
      |   |   |
      H   H   H

      H   OH  OH
      |   |   |
Q  H—C—C—C—H
      |   |   |
      H   H   H

      OH  OH  OH
      |   |   |
R  H—C—C—C—H
      |   |   |
      H   H   H
```

Which line in the table describes the trends in boiling points and viscosities on moving from P to Q to R?

| | *Boiling point* | *Viscosity* |
|---|---|---|
| A | increases | increases |
| B | increases | decreases |
| C | decreases | increases |
| D | decreases | decreases |

**21.** The Williamson synthesis for the preparation of unsymmetrical ethers (ROR') starting with an alcohol and a haloalkane is summarised in the general equations:

Step 1: $ROH + Na \rightarrow RO^{-}Na^{+} + \frac{1}{2}H_2$
Step 2: $RO^{-}Na^{+} + R'X \rightarrow ROR' + Na^{+}X^{-}$

Using propan-2-ol and 2-chlorobutane, the unsymmetrical ether formed would be

A $CH_3CH_2CH_2OCH(CH_3)CH_2CH_3$

B $CH_3CH_2CH_2OCH_2CH_2CH_2CH_3$

C $CH_3CH(CH_3)OCH_2CH_2CH_2CH_3$

D $CH_3CH(CH_3)OCH(CH_3)CH_2CH_3$.

**22.**

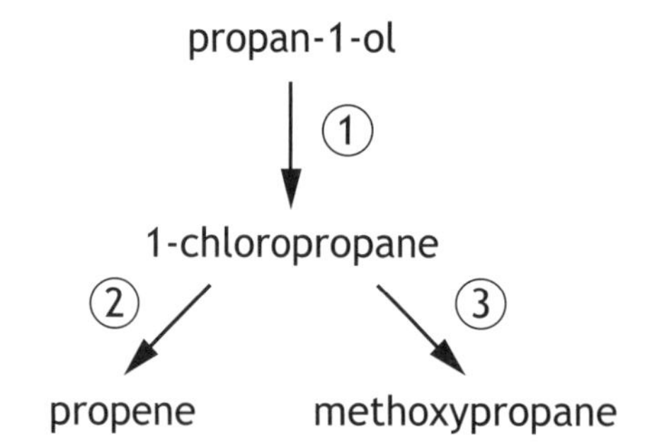


Which line in the table is correct for the types of reaction taking place at ①, ② and ③?

| | *Reaction* ① | *Reaction* ② | *Reaction* ③ |
|---|---|---|---|
| A | substitution | elimination | substitution |
| B | substitution | reduction | substitution |
| C | addition | reduction | condensation |
| D | addition | elimination | substitution |

**23.** P —Cl

Q $CH_2{=}CHCl$

R $CH_2{=}CHCH_2Cl$

Which of the above molecules is/are planar?

A P only

B Q and R only

C P and Q only

D P, Q and R

**24.** One mole of which of the following compounds will react with the largest volume of 1 mol $l^{-1}$ hydrochloric acid?

A HOOC—$C_6H_4$—$NH_2$

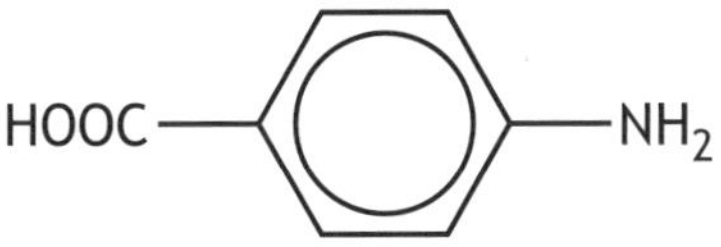

B $HOOCCH_2NH_2$

C $H_2NCH_2NH_2$

D $CH_3NHCH_3$

**25.** Which of the following amines shows no infra-red absorption between 3300 $cm^{-1}$ and 3500 $cm^{-1}$?

A $(CH_3)_3N$

B $CH_3NHCH_3$

C $H_2NCH_2NH_2$

D $C_6H_5$—$NH_2$

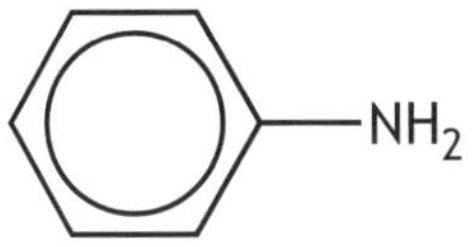

**26.** Which of the following reactions is **least** likely to take place?

A $C_6H_6 \xrightarrow{Br_2/AlCl_3} C_6H_5Br$

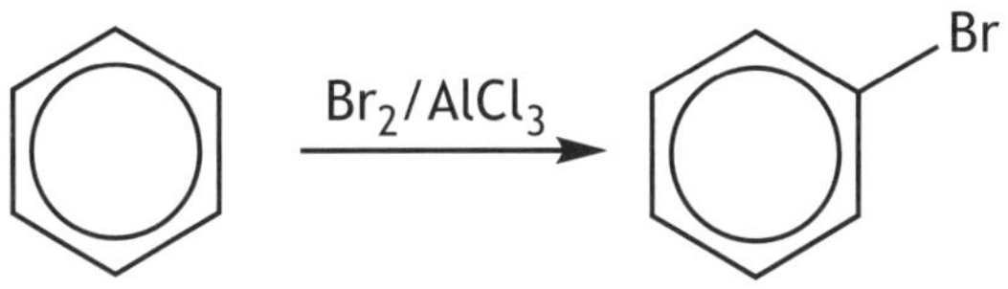

B $C_6H_6 \xrightarrow{Br_2/light} C_6H_5Br$

C $C_6H_5CH_3 \xrightarrow{Br_2/light} C_6H_5CH_2Br$

D $C_6H_6 \xrightarrow{H_2SO_4/SO_3} C_6H_5SO_3H$

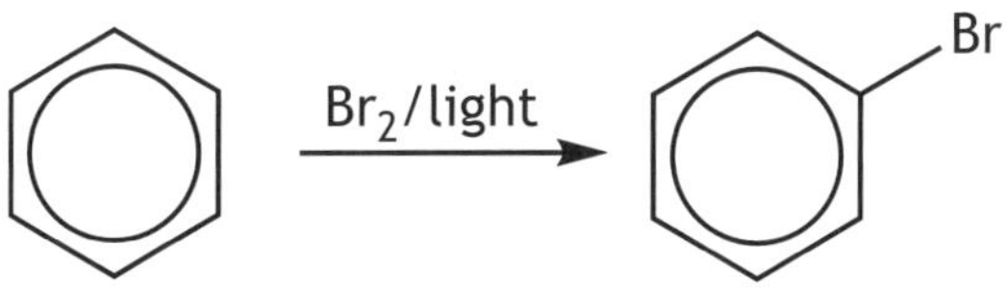

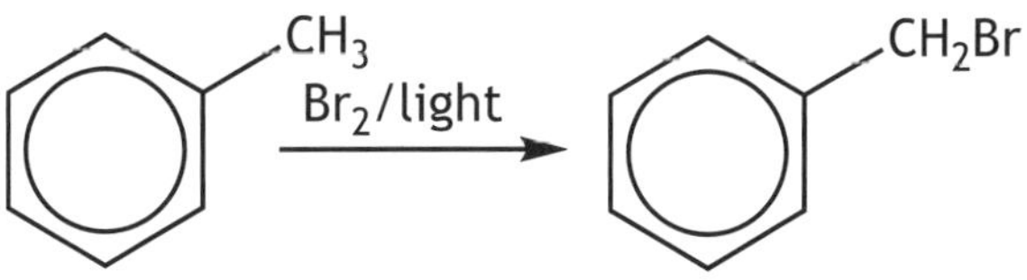

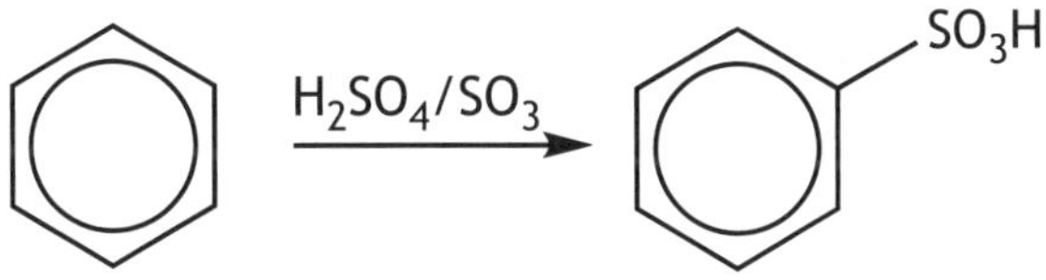

**27.** Which of the following analytical techniques would be most suitable to determine quantitatively the concentration of sodium ions in a urine sample?

A Mass spectrometry

B Infra-red spectroscopy

C Atomic emission spectroscopy

D Proton nuclear magnetic resonance spectroscopy

**28.** The hydrolysis of the haloalkane $(CH_3)_3CBr$ was found to take place by an $S_N1$ mechanism.

The rate determining step involved the formation of

A
$CH_3$
$C^{\oplus}$
$H_3C$ $CH_3$

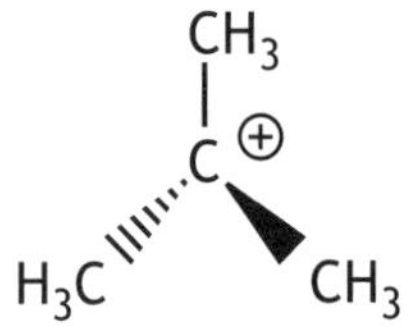

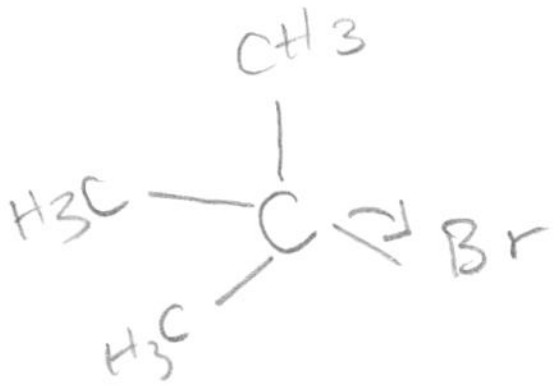

B
$CH_3$
$C^{\ominus}$
$H_3C$ $CH_3$

C
$\left[ \begin{array}{c} CH_3 \\ HO---C---Br \\ H_3C \quad CH_3 \end{array} \right]^{\oplus}$

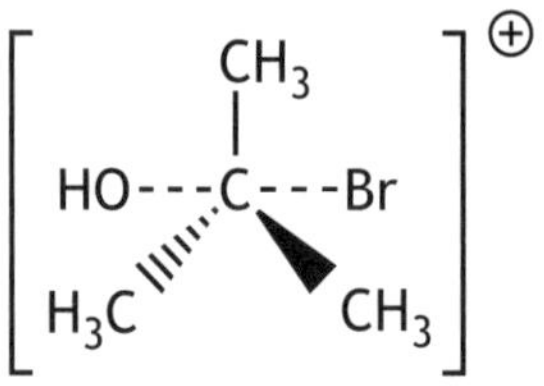

D
$\left[ \begin{array}{c} CH_3 \\ HO---C---Br \\ H_3C \quad CH_3 \end{array} \right]^{\ominus}$

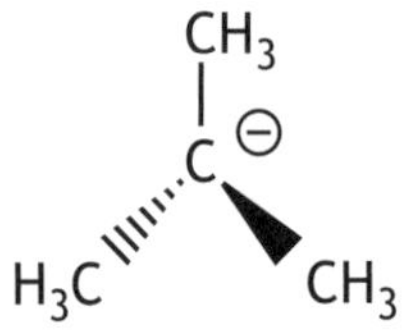

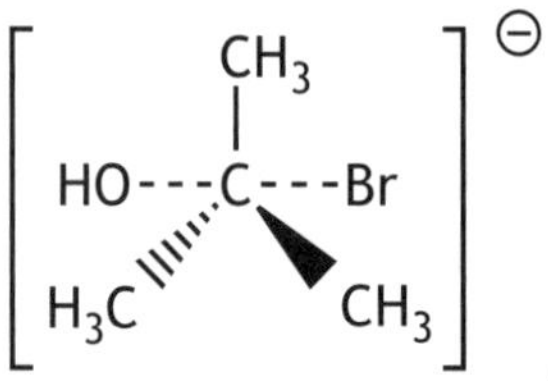

**29.** One of the splitting patterns seen in the high resolution $^{1}H$ NMR spectrum of 3-methylbutan-1-ol below is shown.



**3-methylbutan-1-ol**

Which of the following H atoms, $H_A$, $H_B$, $H_C$ and $H_D$, would produce this splitting pattern?

A $H_A$

B $H_B$

C $H_C$

D $H_D$

**30.** The table shows the structural formulae of some sulfonamides and their antibacterial activity.

| *Sulfonamide* | *Antibacterial activity* |
|---|---|
| 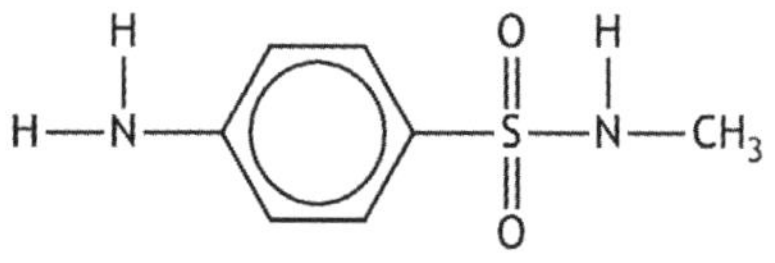 $H_2N-C_6H_4-SO_2-NH-CH_3$ | Active |
| 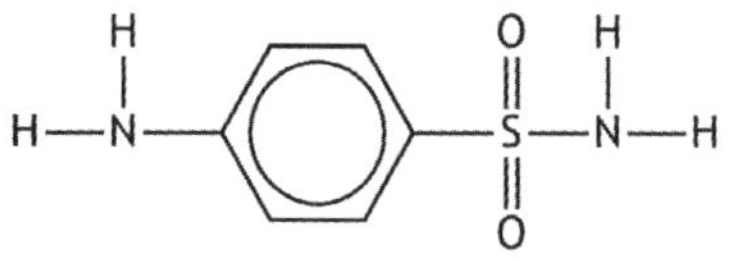 $H_2N-C_6H_4-SO_2-NH-H$ | Active |
| 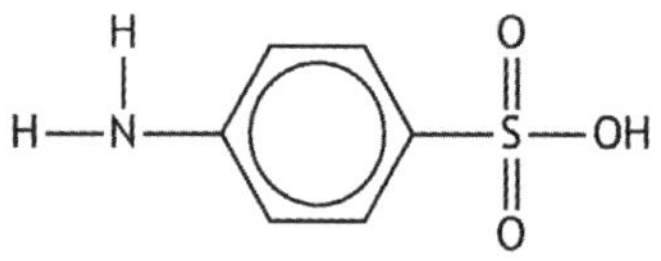 $H_2N-C_6H_4-SO_2-OH$ | Inactive |
| 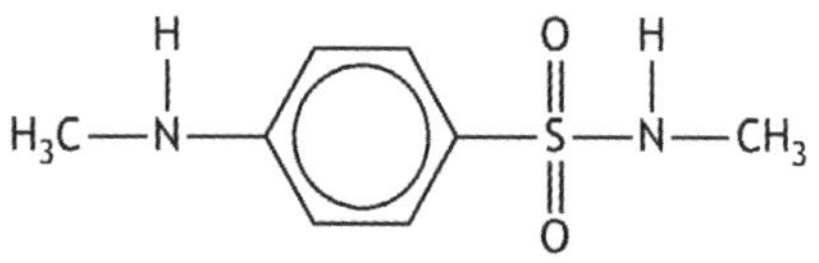 $H_3C-NH-C_6H_4-SO_2-NH-CH_3$ | Inactive |

Which of the following would be an active antibacterial agent?

A $H_2N-C_6H_4-SO_2-NH-$(4,6-dimethylpyridin-2-yl ring: N in ring, two $CH_3$ substituents)

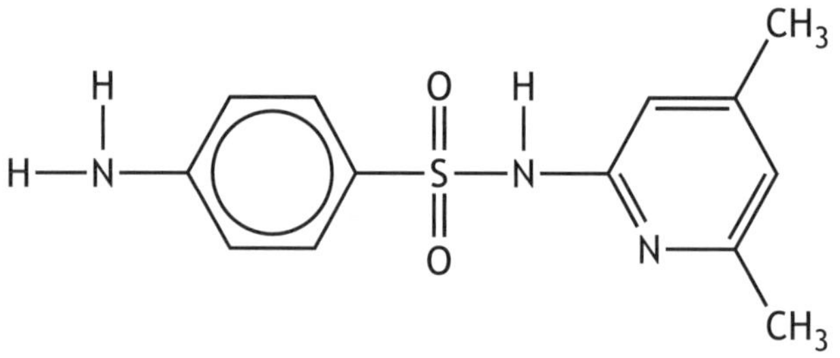

B $H_3C-NH-C_6H_4-SO_2-NH-H$

C $H_2N-C_6H_4-SO_2-Cl$

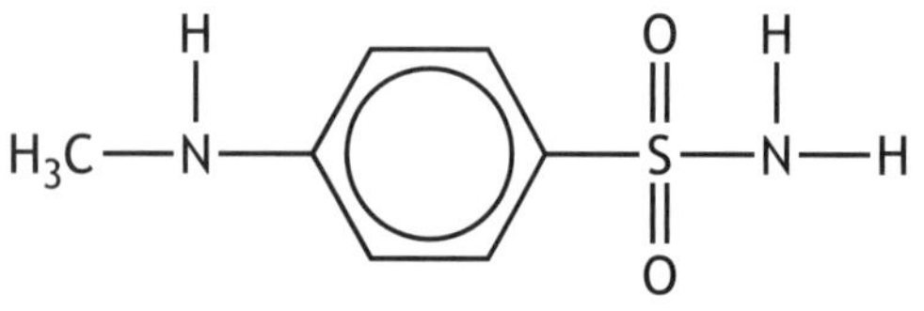

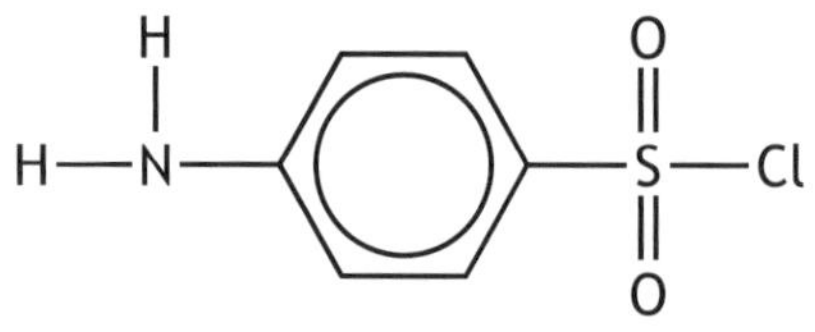

D $H(CH_3)N-C_6H_4-SO_2-NH-$(pyrimidin-2-yl ring: two N in ring)

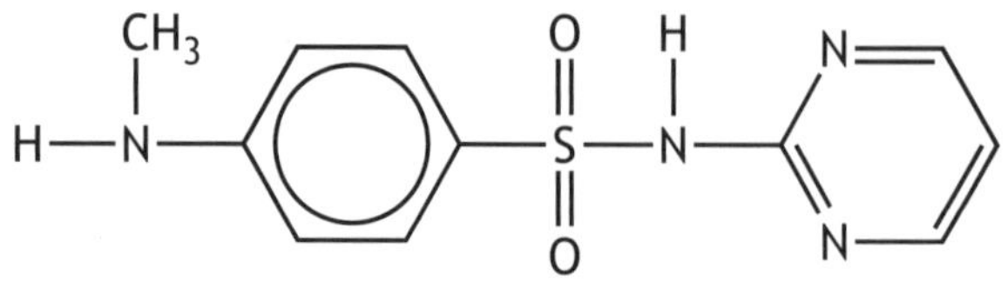

**[END OF SECTION 1. NOW ATTEMPT THE QUESTIONS IN SECTION 2 OF YOUR QUESTION AND ANSWER BOOKLET.]**

FOR OFFICIAL USE

National
Qualifications
SPECIMEN ONLY

Mark

# SQ05/AH/01

# Chemistry Section 1 — Answer Grid and Section 2

Date — Not applicable

Duration — 2 hours 30 minutes

**Fill in these boxes and read what is printed below.**

Full name of centre

Town

Forename(s)

Surname

Number of seat

Date of birth

Day Month Year

D D M M Y Y

Scottish candidate number

Reference may be made to the Chemistry Higher and Advanced Higher Data Booklet.

**Total marks — 100**

**SECTION 1 — 30 marks**

Attempt ALL questions.

Instructions for completion of Section 1 are given on *Page two*.

**SECTION 2 — 70 marks**

Attempt ALL questions

Write your answers clearly in the spaces provided in this booklet. Additional space for answers and rough work is provided at the end of this booklet. If you use this space you must clearly identify the question number you are attempting. Any rough work must be written in this booklet. You should score through your rough work when you have written your final copy.

Use **blue** or **black** ink.

Before leaving the examination room you must give this booklet to the Invigilator; if you do not, you may lose all the marks for this paper.

**SECTION 1— 30 marks**

The questions for Section 1 are contained in the question paper SQ05/AH/02.
Read these and record your answers on the answer grid on *Page three* opposite.
Do NOT use gel pens.

1. The answer to each question is **either** A, B, C or D. Decide what your answer is, then fill in the appropriate bubble (see sample question below).

2. There is **only one correct** answer to each question.

3. Any rough working should be done on the additional space for answers and rough work at the end of this booklet.

**Sample Question**

To show that the ink in a ball-pen consists of a mixture of dyes, the method of separation would be:

A fractional distillation

B chromatography

C fractional crystallisation

D filtration.

The correct answer is **B**—chromatography. The answer **B** bubble has been clearly filled in (see below).

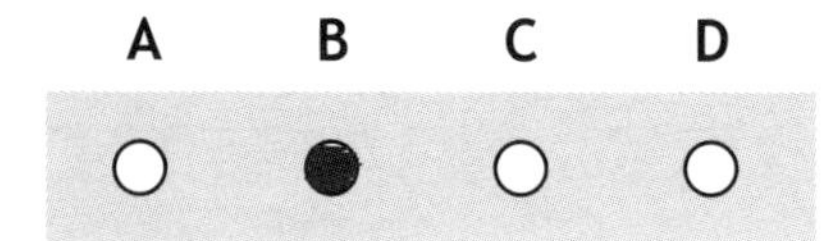


**Changing an answer**

If you decide to change your answer, cancel your first answer by putting a cross through it (see below) and fill in the answer you want. The answer below has been changed to **D**.

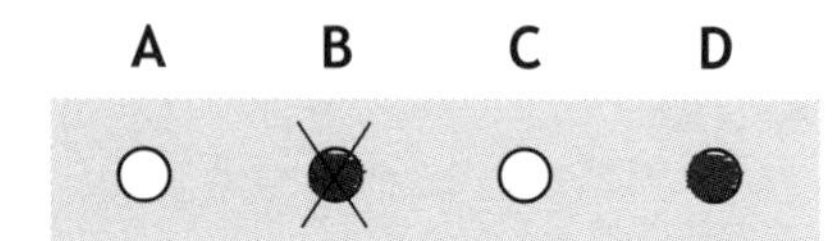


If you then decide to change back to an answer you have already scored out, put a tick (✓) to the **right** of the answer you want, as shown below:

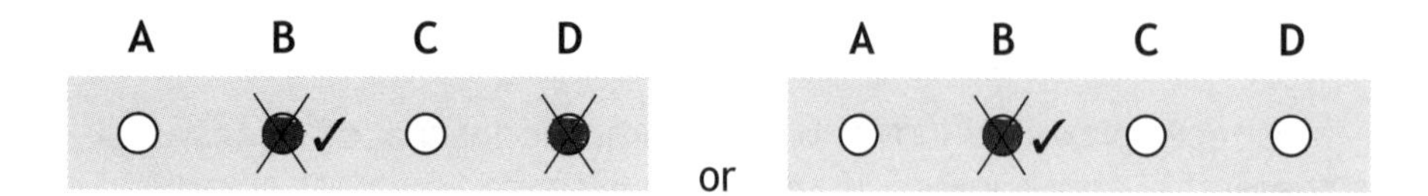

**SECTION 1 — Answer Grid**

| | A | B | C | D |
|---|---|---|---|---|
| 1 | ○ | ○ | ○ | ○ |
| 2 | ○ | ○ | ○ | ○ |
| 3 | ○ | ○ | ○ | ○ |
| 4 | ○ | ○ | ○ | ○ |
| 5 | ○ | ○ | ○ | ○ |
| 6 | ○ | ○ | ○ | ○ |
| 7 | ○ | ○ | ○ | ○ |
| 8 | ○ | ○ | ○ | ○ |
| 9 | ○ | ○ | ○ | ○ |
| 10 | ○ | ○ | ○ | ○ |
| 11 | ○ | ○ | ○ | ○ |
| 12 | ○ | ○ | ○ | ○ |
| 13 | ○ | ○ | ○ | ○ |
| 14 | ○ | ○ | ○ | ○ |
| 15 | ○ | ○ | ○ | ○ |
| 16 | ○ | ○ | ○ | ○ |
| 17 | ○ | ○ | ○ | ○ |
| 18 | ○ | ○ | ○ | ○ |
| 19 | ○ | ○ | ○ | ○ |
| 20 | ○ | ○ | ○ | ○ |
| 21 | ○ | ○ | ○ | ○ |
| 22 | ○ | ○ | ○ | ○ |
| 23 | ○ | ○ | ○ | ○ |
| 24 | ○ | ○ | ○ | ○ |
| 25 | ○ | ○ | ○ | ○ |
| 26 | ○ | ○ | ○ | ○ |
| 27 | ○ | ○ | ○ | ○ |
| 28 | ○ | ○ | ○ | ○ |
| 29 | ○ | ○ | ○ | ○ |
| 30 | ○ | ○ | ○ | ○ |

MARKS | DO NOT WRITE IN THIS MARGIN

## SECTION 2 — 70 marks

### Attempt ALL questions

1. In 2002, astronomers observed a red giant star flash 10 000 times brighter than normal. Its electromagnetic spectrum revealed an intense crimson red line, wavelength 670·7 nm.

   (a) Identify an element present in the red giant star that could be responsible for this intense crimson red line in the emission spectrum. **1**

   (b) Explain how the line of red light is produced. **2**

   (c) Calculate the energy, in $kJ\,mol^{-1}$, associated with this wavelength. **2**

MARKS | DO NOT WRITE IN THIS MARGIN

**2.** Most commercial bleaches contain hypochlorous acid. This acid dissociates as follows:

$$HClO(aq) + H_2O(\ell) \rightleftharpoons H_3O^+(aq) + ClO^-(aq)$$

(a) Identify the conjugate base of hypochlorous acid. **1**

(b) Write the expression for the dissociation constant, $K_a$, for hypochlorous acid. **1**

(c) A solution of hypochlorous acid was titrated with sodium hydroxide solution.

The solution at the end point was alkaline.

Explain why the solution at the end point was alkaline. **2**

MARKS DO NOT WRITE IN THIS MARGIN

**3.** β-carotene is an orange substance and is one of the compounds responsible for the colour of carrots and autumn leaves.

$CH_3$ $CH_3$ $CH_3$ $CH_3$ $H_3C$ $CH_3$ $CH_3$ $CH_3$ $CH_3$ $CH_3$

**β-carotene**

**Using your knowledge of chemistry**, comment on why compounds, such as β-carotene, are the colour they are. **3**

MARKS DO NOT WRITE IN THIS MARGIN

4. The ore pyrolusite, $MnO_2$, was used 30 000 years ago as a black pigment in the cave paintings of Lascaux, France.

The best known oxide of manganese is possibly potassium permanganate, $KMnO_4$, first made in 1740 for the glass industry. It now has many uses including disinfectants and the removal of organic impurities from waste gases and effluent water.

(a) (i) State the oxidation number of manganese in $MnO_2$. 1

(ii) Using orbital box notation, write the electronic configuration for a manganese ion in $MnO_2$. 1

(iii) Explain how your answer is consistent with Hund's rule. 1

(b) (i) The d orbitals in an isolated manganese atom are degenerate. State the meaning of the term degenerate. 1

MARKS | DO NOT WRITE IN THIS MARGIN

**4. (b) (continued)**

(ii) The second quantum number, $\ell$, is related to the shape of the orbitals.

Draw the shape of an orbital when $\ell = 1$. **1**

(c) Small amounts of manganese are added to the aluminium used for drinks cans to improve their corrosion resistance. The technique of colorimetry can be used to determine the quantity of manganese in these alloys and involves converting the manganese to permanganate ions, $MnO_4^-$.

(i) Describe how the technique of colorimetry can be used to determine the concentration of permanganate ions. **3**

(ii) During colorimetric analysis, 0·35 g of an aluminium alloy was dissolved in nitric acid. The manganese in the resulting solution was oxidised and the solution was made up to 250 $cm^3$.

The concentration of this solution was found to be $4{\cdot}25 \times 10^{-4}$ mol $l^{-1}$.

Calculate the percentage, by mass, of manganese in the alloy. **2**

MARKS | DO NOT WRITE IN THIS MARGIN

5. A student was investigating the percentage calcium carbonate content in different types of egg shells. The egg shells were ground and approximately 0·4 g were weighed accurately. The shells were placed in a beaker and 20·0 $cm^3$ of 1·00 mol $l^{-1}$ hydrochloric acid was added. Once the reaction was complete, the solution was made up to 100·0 $cm^3$ in a standard flask.

$$CaCO_3(s) + 2HCl(aq) \rightarrow CaCl_2(aq) + CO_2(g) + H_2O(\ell)$$

(a) State what is meant by weighing accurately, approximately 0·4 g. **1**

(b) Describe the steps required to prepare the 100·0 $cm^3$ solution. **2**

(c) 10·0 $cm^3$ aliquots of the solution were titrated against 0·100 mol $l^{-1}$ standardised sodium hydroxide solution using phenolphthalein as an indicator until concordant results were obtained.

$$NaOH(aq) + HCl(aq) \rightarrow NaCl(aq) + H_2O(\ell)$$

(i) State why the sodium hydroxide solution had to be standardised. **1**

MARKS | DO NOT WRITE IN THIS MARGIN

**5. (c) (continued)**

(ii) An egg shell sample of 0·390 g was used in this experiment.

This led to an average titre volume of 12·65 $cm^3$.

Calculate the percentage, by mass, of calcium carbonate present in the egg shell. **4**

MARKS DO NOT WRITE IN THIS MARGIN

**6.** Omeprazole is a drug commonly used to prevent stomach ulcers. It is described as a proton pump inhibitor as it reduces the ability of enzymes to produce gastric acid. Omeprazole exhibits optical isomerism and is sold as a mixture of both enantiomers.

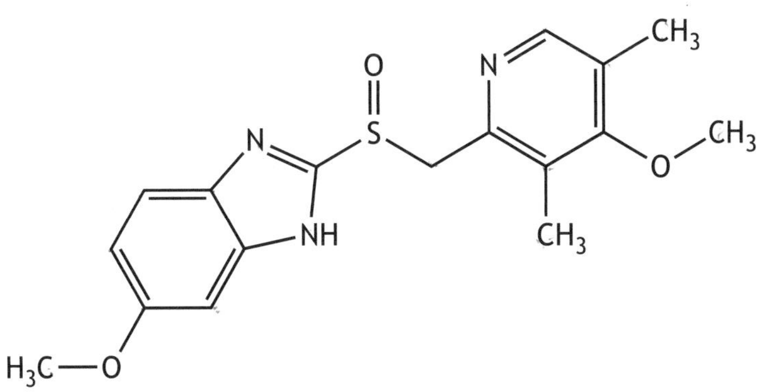

**omeprazole**

(a) Write the molecular formula for omeprazole. **1**

(b) State the name given to a mixture containing equal amounts of both enantiomers. **1**

(c) Suggest the drug classification that best describes omeprazole. **1**

(d) Only one omeprazole enantiomer, known as esomeprazole, is active. However, in acidic environments the other, inactive, enantiomer is converted into the active one.

(i) Explain the benefit of selling the drug as an equal mixture of both enantiomers. **1**

MARKS DO NOT WRITE IN THIS MARGIN

**6. (d) (continued)**

(ii) The first stage of the conversion of the inactive enantiomer involves the reaction with $H^+$ ions as shown.

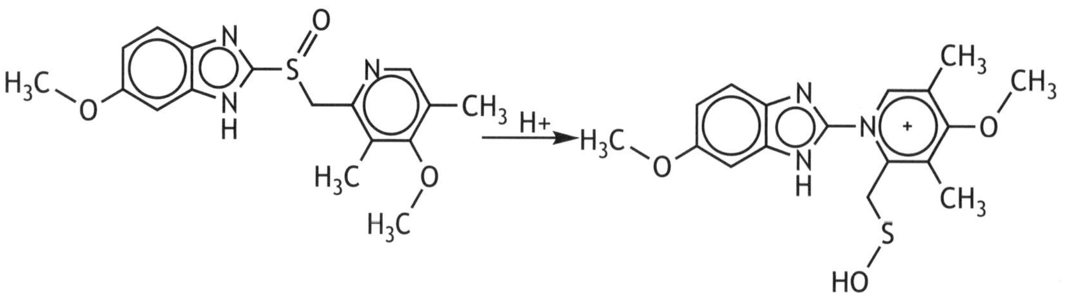

$H_3C$ O N N H S O $CH_3$ $H_3C$ O $H_3C$ H+ $H_3C$ O N N H N + $CH_3$ O $CH_3$ $CH_3$ S HO

**esomeprazole** **compound X**

Identify a functional group that is present in compound X but not in esomeprazole. **1**

MARKS DO NOT WRITE IN THIS MARGIN

**7.** To make nuclear fuel from uranium ore, the element uranium has to be extracted from the ore before being made into fuel pellets.

One of the reactions in the production of nuclear fuel from uranium ore is

$$UO_2(s) + 4HF(g) \rightarrow UF_4(s) + 2H_2O(g) \quad \Delta H^o = -244\,kJ\,mol^{-1}$$

The data in the table below refers to the substances at 298 K.

| *Substance* | $S^o$ / $J\,K^{-1}\,mol^{-1}$ |
|---|---|
| $UO_2(s)$ | 77 |
| $HF(g)$ | 174 |
| $UF_4(s)$ | 152 |
| $H_2O(g)$ | 189 |

(a) Use the data to calculate the entropy change, in $J\,K^{-1}\,mol^{-1}$, at 298 K for this reaction. **2**

(b) Determine, by calculation, whether this reaction is feasible at 298 K. **3**

MARKS DO NOT WRITE IN THIS MARGIN

**8.** Benzocaine is used to relieve pain and itching caused by conditions such as sunburn, insect bites or stings.

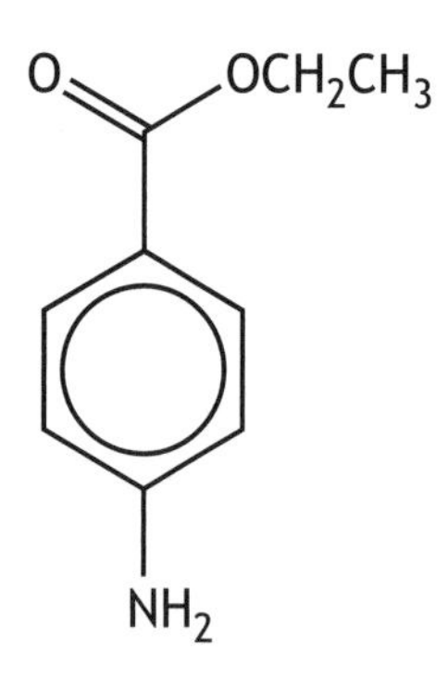


**benzocaine**

A student was carrying out a project to synthesise benzocaine. Part of the procedure to isolate the synthesised benzocaine is given below.

1. Add 20 $cm^3$ of diethyl ether to the reaction mixture and pour into a separating funnel.
2. Add 20 $cm^3$ of distilled water to the separating funnel.
3. Stopper the funnel, invert and gently shake.
4. Allow the aqueous layer to settle to the bottom.

(a) (i) Name the technique described in this part of the procedure. **1**

(ii) Outline the next steps the student will need to carry out in order to obtain a maximum yield of solid benzocaine from the reaction mixture. **3**

(iii) State **two** properties the solvent must have for it to be appropriate for use in this procedure. **2**

MARKS | DO NOT WRITE IN THIS MARGIN

**8. (continued)**

(b) Suggest a second technique that could be used to purify a solid sample of benzocaine. **1**

(c) Thin layer chromatography (TLC) was used to help confirm the identity of the product. A sample of product was dissolved in a small volume of solvent and spotted onto a TLC plate. The plate was allowed to develop and the following chromatogram was obtained.

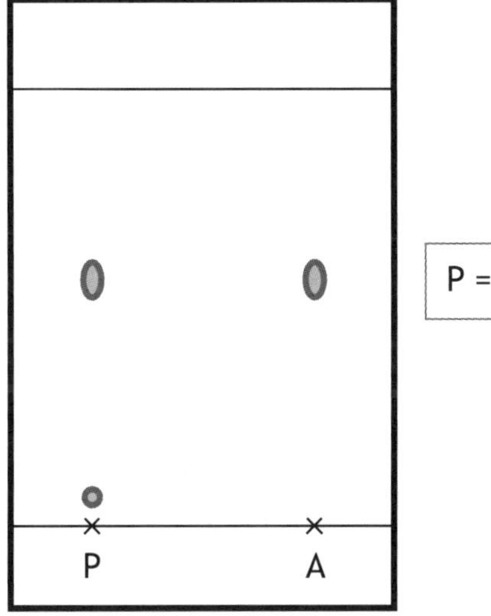


P = student's product

(i) State the name of the substance spotted at A on the TLC plate. **1**

(ii) Evaluate the purity of the student's product. **1**

MARKS DO NOT WRITE IN THIS MARGIN

**9.** At 1000 °C, nitric oxide can combine with hydrogen.

$$2NO(g) + 2H_2(g) \rightarrow N_2(g) + 2H_2O(g)$$

The rate of the above reaction was monitored at different concentrations of NO(g) and $H_2$(g). The results are shown in the table.

| ***Experiment*** | *[NO]* mol $l^{-1}$ | *[$H_2$]* mol $l^{-1}$ | *Initital rate*/mol $l^{-1}$ $s^{-1}$ |
|---|---|---|---|
| 1 | $4{\cdot}00 \times 10^{-3}$ | $1{\cdot}00 \times 10^{-3}$ | $1{\cdot}20 \times 10^{-5}$ |
| 2 | $8{\cdot}00 \times 10^{-3}$ | $1{\cdot}00 \times 10^{-3}$ | $4{\cdot}80 \times 10^{-5}$ |
| 3 | $8{\cdot}00 \times 10^{-3}$ | $4{\cdot}00 \times 10^{-3}$ | $1{\cdot}92 \times 10^{-4}$ |

(a) Determine the order of the reaction with respect to:

(i) NO(g) **1**

(ii) $H_2$(g). **1**

(b) Write the overall rate equation for the reaction. **1**

(c) Calculate a value for the rate constant, k, including the appropriate units. **2**

MARKS DO NOT WRITE IN THIS MARGIN

**10.** Pulegone, first isolated in 1891 from oil of pennyroyal, is a naturally occurring colourless oily liquid with an odour of peppermint. It is classified as a monoterpene and is used in perfumes and aromatherapy.

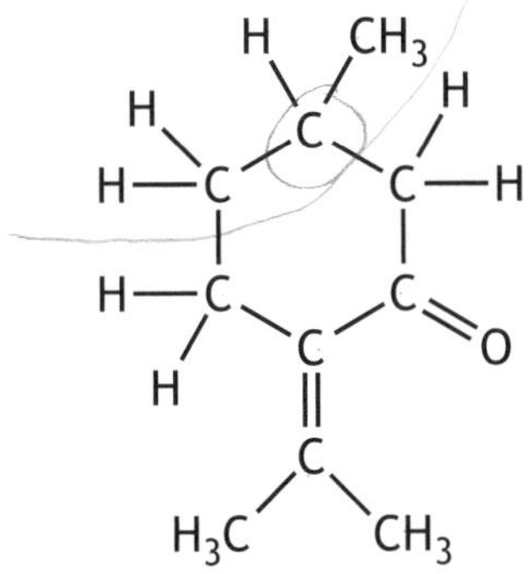

**pulegone**

(a) Circle the chiral centre on the structure of the pulegone above. **1**

(b) Suggest the most appropriate spectroscopic method for identifying the carbonyl group in pulegone. **1**

(c) If pulegone is treated with acid it is converted into two ketones, $C_3H_6O$ and $C_7H_{12}O$.

Draw a possible structure for each of the ketones. **2**

MARKS DO NOT WRITE IN THIS MARGIN

**11.** Cisplatin was the first member of a class of platinum-containing anti-cancer drugs.

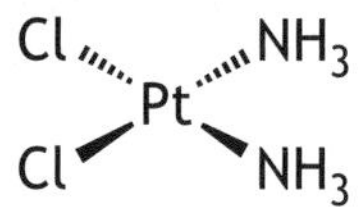

**cisplatin**

Clinical use of the drug is now limited since cancer cells can develop resistance to it.

(a) (i) Explain the meaning of *cis* in cisplatin. **1**

(ii) In the cisplatin complex, chloride ions and ammonia molecules are both classed as monodentate ligands.

Explain the term *monodentate*. **1**

(b) A new drug being trialled, asplatin, is proving to be 10 times more effective than cisplatin.

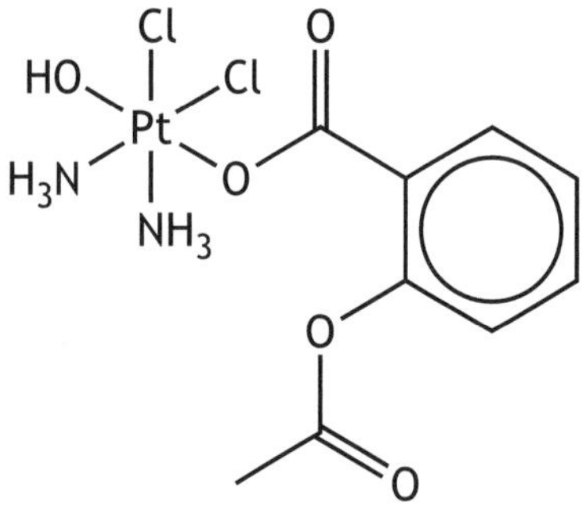

**asplatin**

MARKS | DO NOT WRITE IN THIS MARGIN

**11. (b) (continued)**

Asplatin is synthesised by reacting oxoplatin with acetylsalicylic anhydride.

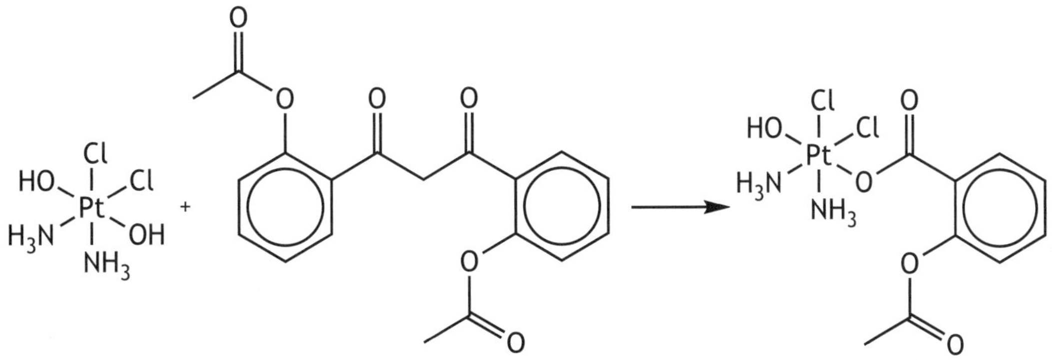

oxoplatin + acetylsalicylic anhydride → asplatin

| oxoplatin | acetylsalicylic anhydride | asplatin |
|---|---|---|
| GFM = 334·1 g | GFM = 342·3 g | GFM = 484·1 g |
| m = 5·00g | | m = 6·36g. |

During a trial synthesis, 5·00 g of oxoplatin was reacted with excess acetylsalicylic anhydride to produce 6·36 g of asplatin.

Calculate the percentage yield. **2**

N = ~~0·0~~ 0·015 moles

N = 0·015 moles

m = 7·2615

$PY = \frac{6 \cdot 36}{7 \cdot 2615} \times 100\%$

= 87·59%

MARKS | DO NOT WRITE IN THIS MARGIN

**12.** Propanoic acid is commonly used in the food industry as a preservative as it can inhibit the growth of mould and bacteria.

Propanoic acid can be prepared from a variety of small molecules. In industry, the main method of production of propanoic acid is by the reaction of ethene with water and carbon monoxide.

$$C_2H_4 + H_2O + CO \rightarrow CH_3CH_2COOH$$

**Using your knowledge of chemistry**, outline the possible steps in the synthesis of propanoic acid from small molecules such as ethene or ethanol. **3**

MARKS DO NOT WRITE IN THIS MARGIN

**13.** Both spectroscopic and chemical analysis can be used to determine the identity of an unknown compound.

The following spectra and chemical data were obtained for a colourless liquid with a pleasant smell.

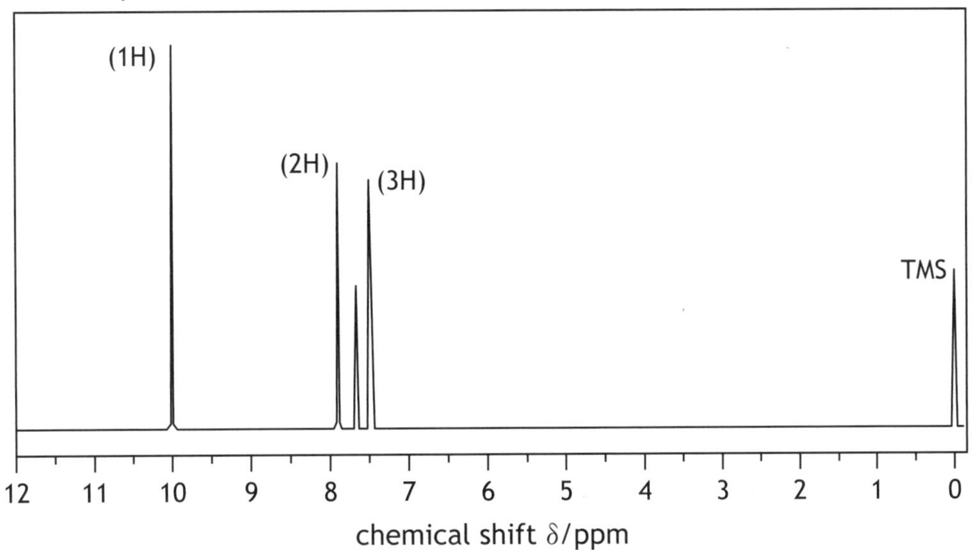

MARKS | DO NOT WRITE IN THIS MARGIN

**13. (continued)**

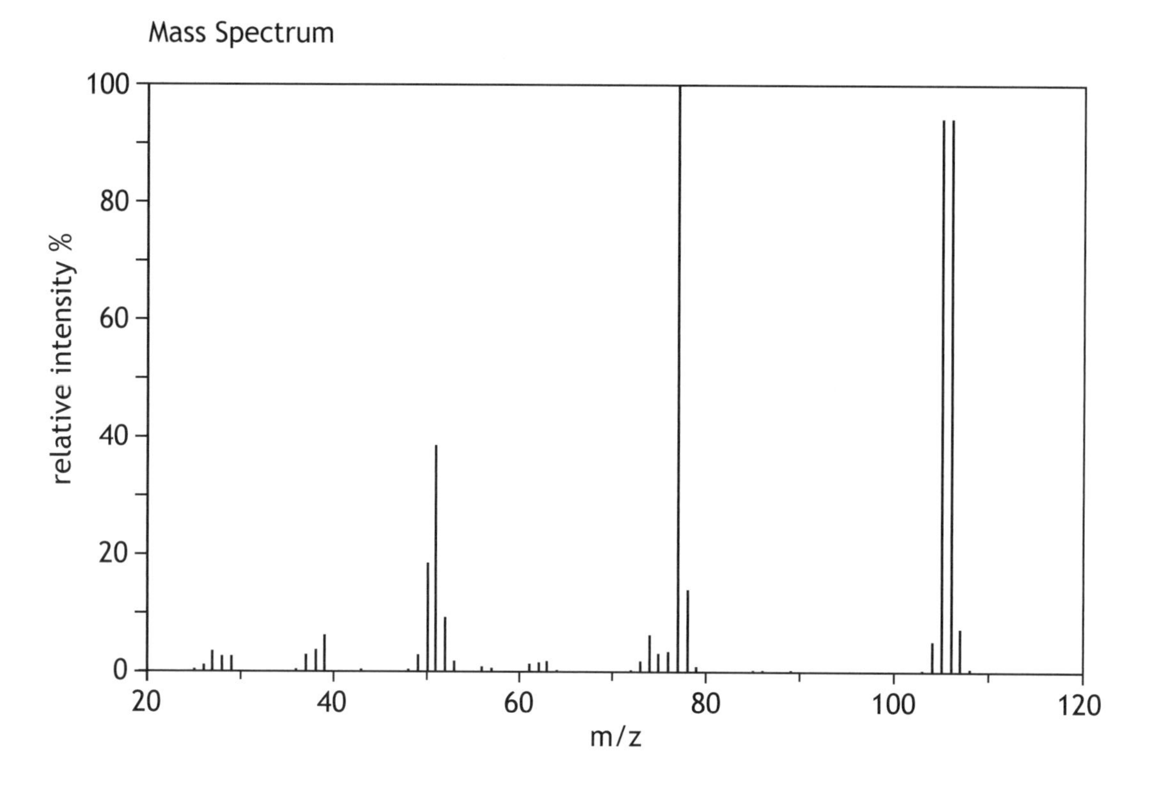


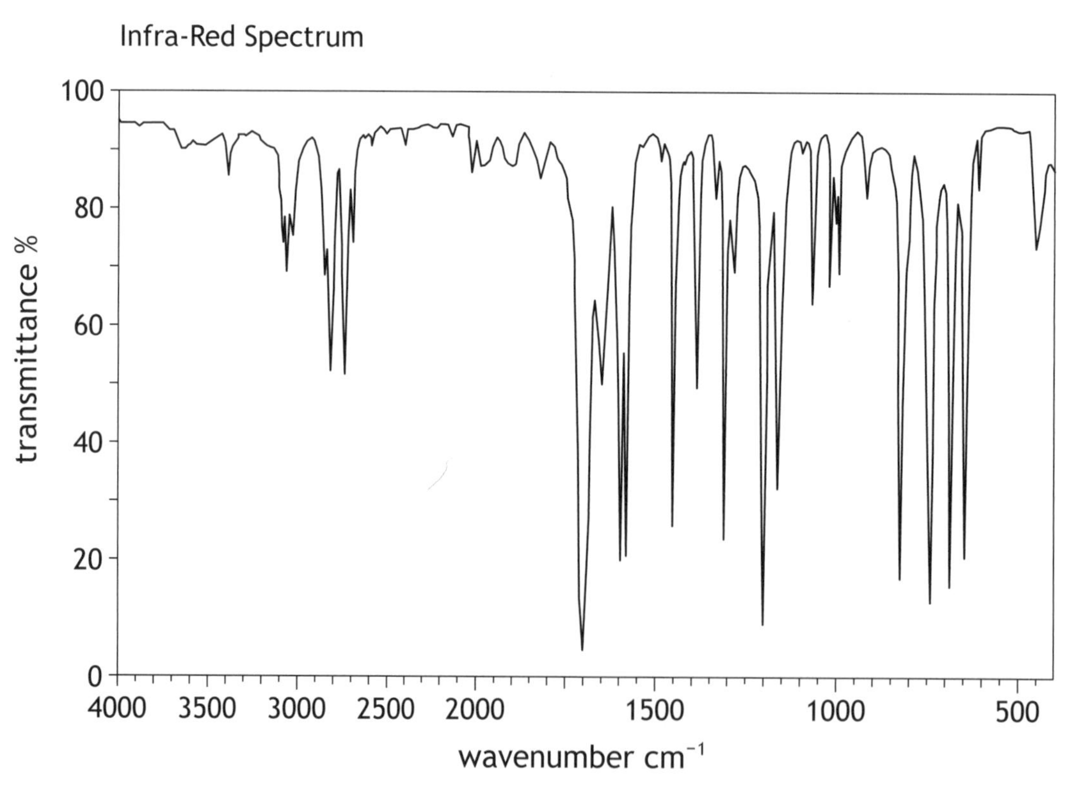


**Chemical Data**

| | |
|---|---|
| Composition: | C 79·25%; H 5·66%; O 15·09% |
| Fehling's/Benedict's: | brick red precipitate forms |
| Bromine solution: | no change |
| Aqueous sodium carbonate: | no reaction |

MARKS | DO NOT WRITE IN THIS MARGIN

**13. (continued)**

(a) (i) Determine the empirical formula for the compound. **2**

(ii) Using the mass spectrum, determine the molecular formula for the compound. **1**

(b) Identify the functional group in the compound which is responsible for the peak at 1700 $cm^{-1}$ in the infra-red spectrum. **1**

(c) Using all of the information, draw a structural formula for the compound. **1**

**[END OF SPECIMEN QUESTION PAPER]**

ADDITIONAL SPACE FOR ANSWERS AND ROUGH WORK

**ADDITIONAL SPACE FOR ANSWERS AND ROUGH WORK**

[BLANK PAGE]

DO NOT WRITE ON THIS PAGE

ADVANCED HIGHER

# Model Paper 1

Whilst this Model Paper has been specially commissioned by Hodder Gibson for use as practice for the Advanced Higher (for Curriculum for Excellence) exams, the key reference document remains the SQA Specimen Paper 2015 and SQA Past Paper 2016.

National
Qualifications
MODEL PAPER 1

# Chemistry
# Section 1 — Questions

Duration — 2 hours 30 minutes

Instructions for the completion of Section 1 are given on *Page two* of your question and answer booklet.

Record your answers on the answer grid on *Page three* of your question and answer booklet.

Reference may be made to the Chemistry Higher and Advanced Higher Data Booklet.

Before leaving the examination room you must give your question and answer booklet to the Invigilator; if you do not you may lose all the marks for this paper.

**SECTION 1 — 30 marks**

**Attempt ALL questions**

**1.** The diagram shows one of the series of lines in the hydrogen emission spectrum.

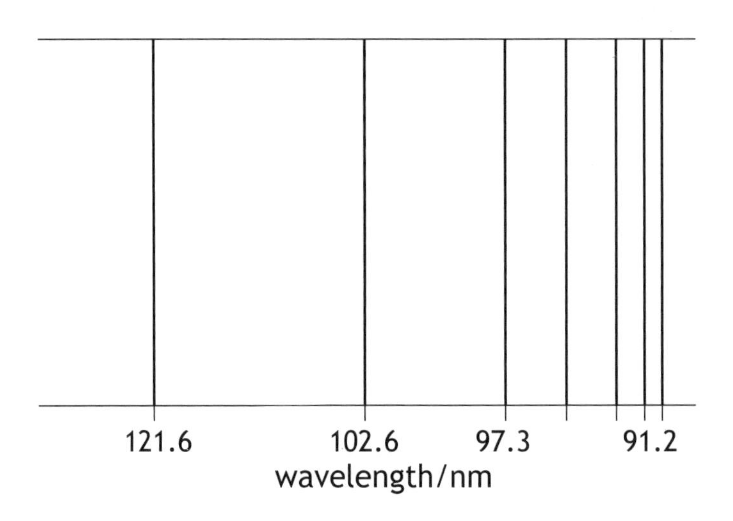


Each line

A represents an energy level within a hydrogen atom

B results from an electron moving to a higher energy level

C lies within the visible part of the electromagnetic spectrum

D results from an excited electron dropping to a lower energy level.

**2.** An atom has electronic configuration

$1s^2\ 2s^2\ 2p^6\ 3s^2\ 3p^3$.

When ionised, this is likely to form a particle with a charge of

A 5+

B 5−

C 3+

D 3−.

**3.** Which line in the table correctly describes the arrangement of electron pairs in $PF_3$ and $PF_5$?

| | $PF_3$ | $PF_5$ |
|---|---|---|
| **A** | Trigonal planar | Trigonal bipyramidal |
| **B** | Tetrahedral | Octahedral |
| **C** | Tetrahedral | Trigonal bipyramidal |
| **D** | Trigonal planar | Octahedral |

**4.** The number of unpaired electrons in a vanadium atom is

A 0

B 2

C 3

D 5.

**5.** At a particular temperature, 8·0 mole of $NO_2$ was placed in a 1 litre container and the $NO_2$ dissociated by the following reaction:

$$2NO_2(g) \rightleftharpoons 2NO(g) + O_2(g)$$

At equilibrium the concentration of NO(g) is 2·0 mol $l^{-1}$.

The equilibrium constant will have a value of

A 0·11

B 0·22

C 0·33

D 9·00.

**6.** The pH of 0·025 mol $l^{-1}$ $H_2SO_4(aq)$ is

A 1·3

B −1·6

C 1·6

D 1·9.

**7.** Ammonia, $NH_3$, can react with hydrogen chloride gas to form the salt ammonium chloride.

$NH_3(g) + HCl(g) \rightarrow NH_4Cl(s)$

In this reaction, ammonia is acting as

A a proton donor

B an acid

C a proton acceptor

D a conjugate base.

**8.** Which of the following would not act as a buffer solution?

A Ethanoic acid and potassium ethanoate

B Nitric acid and potassium nitrate

C Ammonia and ammonium nitrate

D Hexanoic acid and magnesium hexanoate

**9.** Which of the following indicators should be used in the titration of methanoic acid with sodium hydroxide?

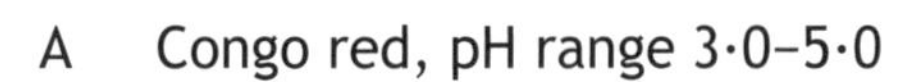

A Congo red, pH range 3·0–5·0

B Methyl red, pH range 4·2–6·2

C Phenolphthalein, pH range 8·0–9·8

D Bromothymol blue, pH range 6·0–7·6

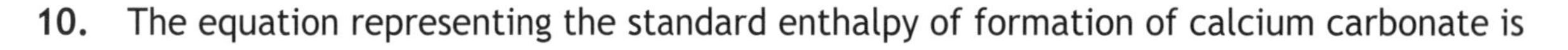

**10.** The equation representing the standard enthalpy of formation of calcium carbonate is

A $Ca(s) + C(s) + 1\frac{1}{2}O_2(g) \rightarrow CaCO_3(s)$

B $Ca(g) + C(g) + 1\frac{1}{2}O_2(g) \rightarrow CaCO_3(g)$

C $Ca(s) + CO_2(g) + \frac{1}{2}O_2(g) \rightarrow CaCO_3(s)$

D $Ca(s) + C(s) + 3O(g) \rightarrow 2CaCO_3(s)$

**11.** Which of the following reactions will have a positive $\Delta S°$ value?

A $C_2H_4(g) + H_2(g) \rightarrow C_2H_6(g)$

B $2K(s) + 2H_2O(l) \rightarrow 2KOH(aq) + H_2(g)$

C $Pb(NO_3)_2(aq) + 2KI(aq) \rightarrow PbI_2(s) + 2KNO_3(aq)$

D $2H_2(g) + O_2(g) \rightarrow 2H_2O(l)$

**12.** The $\Delta G°$ at 298 K for a chemical reaction with $\Delta H° = 10\,kJ\,mol^{-1}$ and $\Delta S° = -14\,J\,K^{-1}\,mol^{-1}$ is

A $14\,kJ\,mol^{-1}$

B $-14\,kJ\,mol^{-1}$

C $-4162\,kJ\,mol^{-1}$

D $4162\,kJ\,mol^{-1}$.

**13.** The following data refer to initial reaction rates obtained for the reaction

X + Y + Z $\rightarrow$ products

| *Run* | *Relative concentrations* | | | *Relative initial rate* |
|---|---|---|---|---|
| | *[X]* | *[Y]* | *[Z]* | |
| 1 | 1·0 | 1·0 | 1·0 | 0·3 |
| 2 | 1·0 | 2·0 | 1·0 | 0·6 |
| 3 | 2·0 | 2·0 | 1·0 | 1·2 |
| 4 | 2·0 | 1·0 | 2·0 | 0·6 |

These data fit the rate equation

A Rate = k[X]

B Rate = k[X][Y]

C Rate = $k[X][Y]^2$

D Rate = k[X][Y][Z].

**14.**

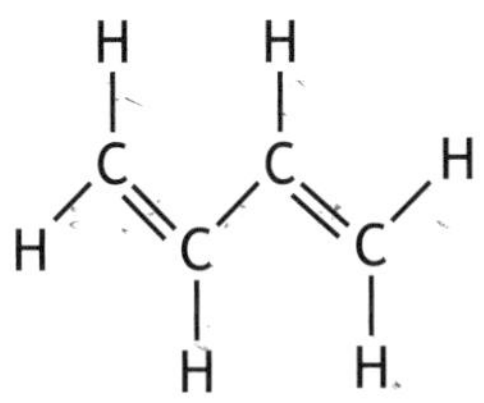

The correct number of sigma and pi bonds in the alkene shown above is

A nine sigma and two pi

B seven sigma and four pi

C eleven sigma and two pi

D eleven sigma and four pi.

**15.** The sideways overlap of the two p orbitals shown below will result in the formation of a

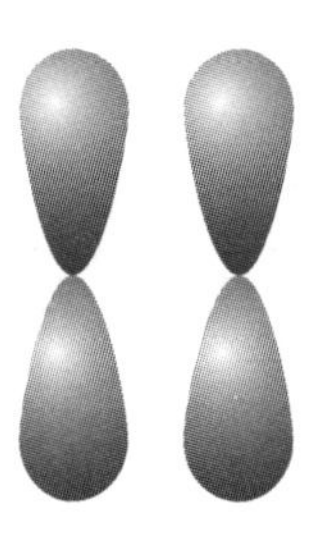

A sigma bond

B pi bond

C hybrid orbital

D sigma bond and a pi bond.

**16.** Which of the following has a geometric isomer?

| | |
|---|---|
| **A** | H H<br>\| \|<br>H—C—C—H<br>\| \|<br>Cl Cl |
| **B** | H H<br>\ /<br>C=C<br>/ \<br>Cl Cl |
| **C** | Cl Cl<br>\ /<br>C=C<br>/ \<br>Cl Cl |
| **D** | Cl Cl<br>\| \|<br>H—C—C—H<br>\| \|<br>Cl Cl |

**17.** Propene can react with hydrogen chloride to form 2-chloropropane. Which line in the table correctly describes this reaction?

| | *Propene is acting as* | *Type of bond fission* |
|---|---|---|
| **A** | a nucleophile | heterolytic |
| **B** | an electrophile | homolytic |
| **C** | a nucleophile | homolytic |
| **D** | an electrophile | heterolytic |

**18.** Alkenes react with ozone ($O_3$) to form ozonides which can be hydrolysed to give carbonyl compounds.

$$\rangle C{=}C\langle \xrightarrow{O_3} \text{OZONIDE} \xrightarrow{\text{hydrolysis}} \rangle C{=}O + O{=}C\langle$$

Which of the following alkenes will produce a mixture of propanone and ethanal when acted upon in this way?

A $CH_3CH{=}CHCH_2CH_3$

B $CH_3CH{=}CHCH_3$

C $(CH_3)_2C{=}CH_2$

D $CH_3CH{=}C(CH_3)_2$

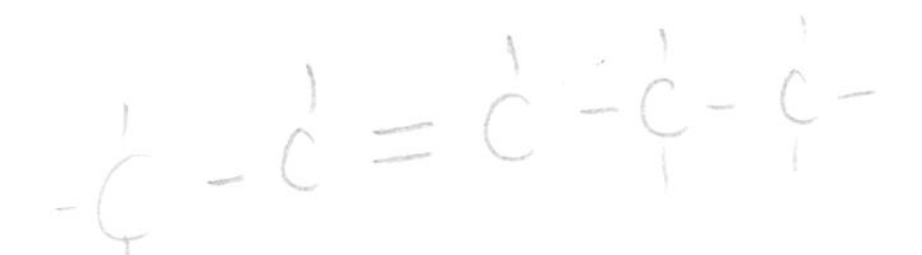

**19.** Which line in the table below correctly describes the reaction of ethanal with LiAlH4?

| | *$LiAlH_4$ acts as* | *Product formed* |
|---|---|---|
| **A** | a reducing agent | ethanoic acid |
| **B** | an oxidising agent | ethanol |
| **C** | a reducing agent | ethanol |
| **D** | an oxidising agent | ethanoic acid |

**20.** Which of the following compounds would produce ethyl propanoate when reacted with ethanol?

| | |
|---|---|
| A | $CH_3CH_2{-}C({=}O){-}H$ |
| B | $CH_3CH_2{-}C({=}O){-}Cl$ |
| C | $CH_3{-}C({=}O){-}CH_3$ |
| D | $CH_3{-}CH(OH){-}CH_3$ |

**21.** Which line in the table correctly describes the reaction of benzene with a mixture of concentrated nitric acid and concentrated sulfuric acid?

| | *Type of reaction* | *Product formed* |
|---|---|---|
| **A** | electrophilic substitution | benzene ring with $NO_2$ |
| **B** | electrophilic substitution | benzene ring with $SO_3H$ |
| **C** | nucleophilic substitution | benzene ring with $NO_2$ |
| **D** | nucleophilic substitution | benzene ring with $SO_3H$ |

**22.** Which of the following cannot be applied to benzene?

A $sp^2$ hybridised carbon atoms

B Takes part in addition reactions

C p orbitals overlap to produce a delocalised ring of electrons

D Planar molecule

**23.**

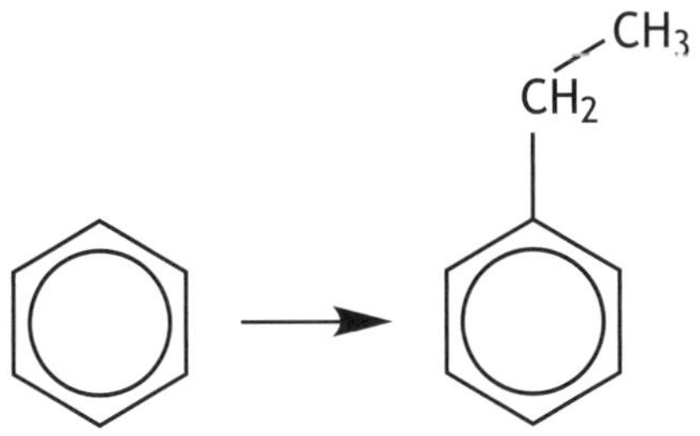

$CH_3$
$CH_2$

The reagents required for the above reaction are

A $CH_3CH_2Cl$

B $CH_3CH_3$ and UV light

C $CH_3CH_2Cl$ and $AlCl_3$

D $CH_3CH_2^-$.

**24.** Which of the following compounds is most likely to show peaks on a mass spectrum at m/z 45 and m/z 31?

A Ethanal

B Ethane

C Ethanoic acid

D Ethanol

**25.** Which of the following spectroscopic techniques involves vibration of chemical bonds?

A Mass spectrometry

B Infrared spectroscopy

C Atomic emission spectroscopy

D Proton nuclear magnetic resonance spectroscopy

**26.** Which of the following compounds is likely to produce the proton NMR spectrum shown below?

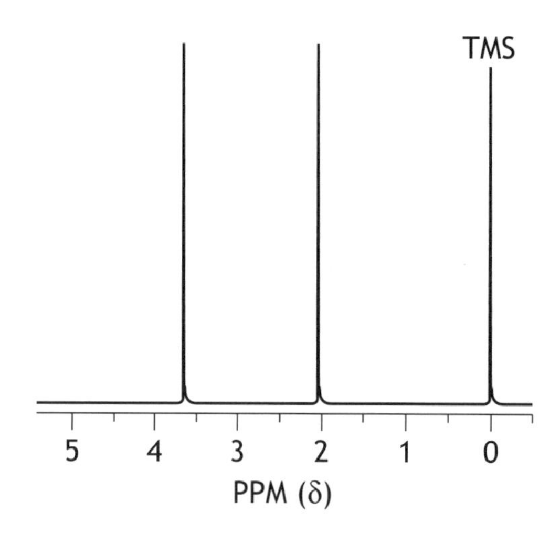


| | |
|---|---|
| A | $CH_3$–$CH_3$ (H–C–C–H with three H on each C) |
| B | $CH_3$–$CH_2$–O–H |
| C | $CH_3$–C(=O)–O–$CH_3$ |
| D | $CH_3$–C(=O)–H |

**27.** Ipratropium is the active ingredient in a medicine used to relieve asthma. The medicine works by binding to a receptor in the lungs causing a specific reaction in the lungs to stop. This allows the patient to breathe more easily.

Which line in the table correctly describes how this medicine works?

| | *Ipratropium acts as an* | *Ipratropium binds to* |
|---|---|---|
| A | agonist | protein |
| B | antagonist | protein |
| C | agonist | fat |
| D | antagonist | fat |

**28.** The most appropriate pieces of equipment to use when diluting a solution by a factor of 10 would be

A a 10·0 $cm^3$ pipette and a 100 $cm^3$ measuring cylinder

B a 10·0 $cm^3$ pipette and a 50 $cm^3$ standard flask

C a 25·0 $cm^3$ measuring cylinder and a 250 $cm^3$ standard flask

D a 25·0 $cm^3$ pipette and a 250 $cm^3$ standard flask.

**29.** Using thin-layer chromatography the components of a mixture can be identified by their $R_f$ values.

Which of the following statements is **true** about the $R_f$ value of an individual component of a mixture?

A The type of stationary phase has no effect on the $R_f$ value.

B The polarity of the component has no effect on the $R_f$ value.

C The composition of the mobile phase has no effect on the $R_f$ value.

D The distance the solvent front moves has no effect on the $R_f$ value.

**30.** An excess of sodium sulfate was added to a solution of a barium compound to precipitate all the barium ions as barium sulfate, $BaSO_4$. (GFM of $BaSO_4$ = 233·4 g.)

How many grams of barium are in 0·458 g of the barium compound if a solution of this sample gave 0·513 g of $BaSO_4$ precipitate?

A 0·032 g

B 0·055 g

C 0·269 g

D 0·302 g

**[END OF SECTION 1. NOW ATTEMPT THE QUESTIONS IN SECTION 2 OF YOUR QUESTION AND ANSWER BOOKLET.]**

[BLANK PAGE]

DO NOT WRITE ON THIS PAGE

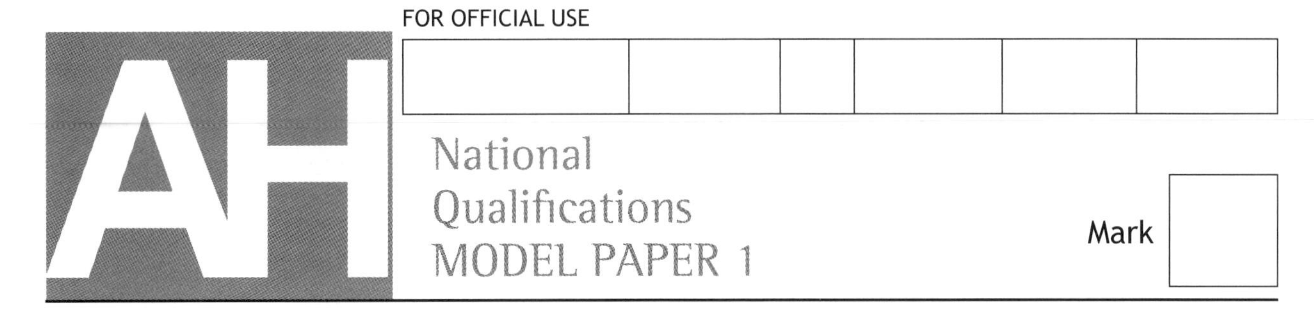


# Chemistry
# Section 1 — Answer Grid and Section 2

Duration — 2 hours 30 minutes

**Fill in these boxes and read what is printed below.**

Full name of centre

Town

Forename(s)

Surname

Number of seat

Date of birth

Day Month Year

D D M M Y Y

Scottish candidate number

Reference may be made to the Chemistry Higher and Advanced Higher Data Booklet.

**Total marks — 100**

**SECTION 1 — 30 marks**

Attempt ALL questions.

Instructions for completion of Section 1 are given on *Page two*.

**SECTION 2 — 70 marks**

Attempt ALL questions

Write your answers clearly in the spaces provided in this booklet. Additional space for answers and rough work is provided at the end of this booklet. If you use this space you must clearly identify the question number you are attempting. Any rough work must be written in this booklet. You should score through your rough work when you have written your final copy.

Use **blue** or **black** ink.

Before leaving the examination room you must give this booklet to the Invigilator; if you do not, you may lose all the marks for this paper.

**SECTION 1 — 30 marks**

The questions for Section 1 are contained in the question paper on *Page four*.
Read these and record your answers on the answer grid on *Page three* opposite.
Do NOT use gel pens.

1. The answer to each question is **either** A, B, C or D. Decide what your answer is, then fill in the appropriate bubble (see sample question below).

2. There is **only one correct** answer to each question.

3. Any rough working should be done on the additional space for answers and rough work at the end of this booklet.

**Sample Question**

To show that the ink in a ball-pen consists of a mixture of dyes, the method of separation would be:

A fractional distillation

B chromatography

C fractional crystallisation

D filtration.

The correct answer is **B**—chromatography. The answer **B** bubble has been clearly filled in (see below).

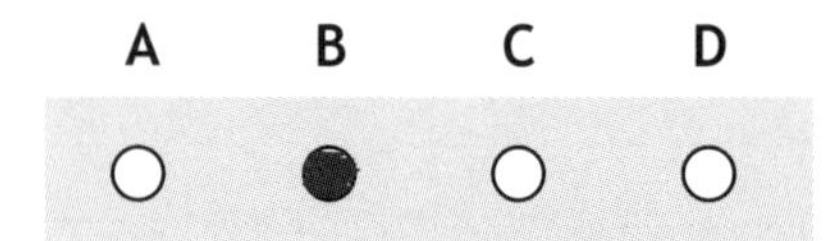


**Changing an answer**

If you decide to change your answer, cancel your first answer by putting a cross through it (see below) and fill in the answer you want. The answer below has been changed to **D**.

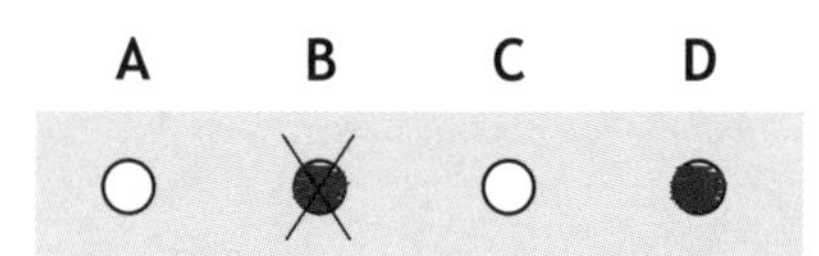


If you then decide to change back to an answer you have already scored out, put a tick (✓) to the **right** of the answer you want, as shown below:

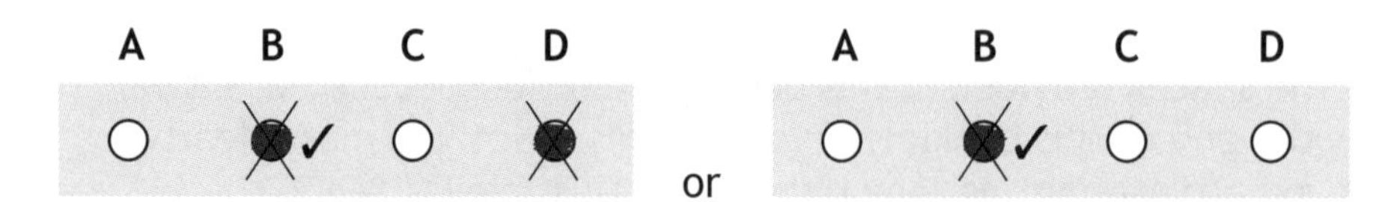

**SECTION 1 — Answer Grid**

| | A | B | C | D |
|---|---|---|---|---|
| 1 | ○ | ○ | ○ | ○ |
| 2 | ○ | ○ | ○ | ○ |
| 3 | ○ | ○ | ○ | ○ |
| 4 | ○ | ○ | ○ | ○ |
| 5 | ○ | ○ | ○ | ○ |
| 6 | ○ | ○ | ○ | ○ |
| 7 | ○ | ○ | ○ | ○ |
| 8 | ○ | ○ | ○ | ○ |
| 9 | ○ | ○ | ○ | ○ |
| 10 | ○ | ○ | ○ | ○ |
| 11 | ○ | ○ | ○ | ○ |
| 12 | ○ | ○ | ○ | ○ |
| 13 | ○ | ○ | ○ | ○ |
| 14 | ○ | ○ | ○ | ○ |
| 15 | ○ | ○ | ○ | ○ |
| 16 | ○ | ○ | ○ | ○ |
| 17 | ○ | ○ | ○ | ○ |
| 18 | ○ | ○ | ○ | ○ |
| 19 | ○ | ○ | ○ | ○ |
| 20 | ○ | ○ | ○ | ○ |
| 21 | ○ | ○ | ○ | ○ |
| 22 | ○ | ○ | ○ | ○ |
| 23 | ○ | ○ | ○ | ○ |
| 24 | ○ | ○ | ○ | ○ |
| 25 | ○ | ○ | ○ | ○ |
| 26 | ○ | ○ | ○ | ○ |
| 27 | ○ | ○ | ○ | ○ |
| 28 | ○ | ○ | ○ | ○ |
| 29 | ○ | ○ | ○ | ○ |
| 30 | ○ | ○ | ○ | ○ |

## SECTION 2 — 70 marks

**Attempt ALL questions**

MARKS DO NOT WRITE IN THIS MARGIN

1. When a high voltage is applied to a lamp filled with neon gas, a line emission spectrum is produced.

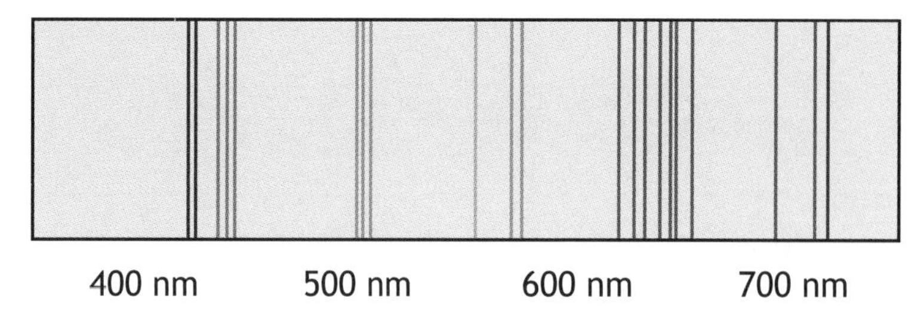


(a) Explain how a line in the line emission spectrum is produced. **2**

(b) Calculate the energy, in $kJ\,mol^{-1}$, associated with the line at 585 nm. **2**

(c) A sample of dry air was found to contain 18 ppm neon. Calculate the volume of neon present in a 10 kg sample of dry air assuming that the molar volume was 22·4 litres $mol^{-1}$. **3**

MARKS | DO NOT WRITE IN THIS MARGIN

**2.** (a) A salt of methanoic acid, sodium methanoate is commonly used as a food preservative.

Explain why sodium methanoate solution has a pH greater than 7.

In your answer you should mention the two equilibria involved. **2**

(b) When sodium m ethanoate is added to methanoic acid, a buffer solution is formed.

0·50 moles of sodium methanoate was added to 200 $cm^3$ of a 0·25 $mol\,l^{-1}$ solution of methanoic acid at 298 K, $pK_a$ = 3·75.

Calculate the pH of the buffer solution formed. **2**

MARKS DO NOT WRITE IN THIS MARGIN

**3.** A student stated that, "Ionisation energy always increases across a period in the periodic table due to the increase in nuclear charge."

Using your knowledge of chemistry, discuss the accuracy of this statement. **3**

MARKS DO NOT WRITE IN THIS MARGIN

**4.** As part of an investigation, a student was analysing the metallic content of a key known to be composed of a copper/nickel alloy.

The key was dissolved in nitric acid and the resulting solution diluted to 1000 $cm^3$ in a standard flask using tap water. Three 25·0 $cm^3$ samples of the nitrate solution were pipetted into separate conical flasks and approximately 10 g of solid potassium iodide were added. Iodine was produced as shown in the equation.

$2Cu^{2+}(aq) + 4I^{-}(aq) \rightarrow 2CuI(s) + I_2(aq)$

The liberated iodine was titrated against standardised 0·102 mol $l^{-1}$ sodium thiosulfate solution. Starch indicator was added near the end point of the titration.

$I_2(aq) + 2S_2O_3^{2-}(aq) \rightarrow 2I^{-}(aq) + S_4O_6^{2-}(aq)$

The results, for the volume of thiosulfate used, are given in the table.

| | *Titration 1* | *Titration 2* | *Titration 3* |
|---|---|---|---|
| **Final volume/$cm^3$** | 16·30 | 31·50 | 46·80 |
| **Initial volume/$cm^3$** | 0·30 | 16·30 | 31·50 |
| **Volume added/$cm^3$** | 16·00 | 15·20 | 15·30 |

(a) From the results calculate the mass of copper in the key. **3**

(b) Suggest how the accuracy of the analysis could be improved. **1**

MARKS | DO NOT WRITE IN THIS MARGIN

**4. (continued)**

(c) When copper(II) is dissolved in water, a light blue coloured solution is formed which has the formula $[Cu(H_2O)_6]^{2+}$. Reaction of this solution with excess ammonia results in the formation of the complex $[Cu(NH_3)_4(H_2O)_2]^{2+}$ which has a deep blue colour.

(i) State the name of the complex ion $[Cu(H_2O)_6]^{2+}$. **1**

(ii) Draw a structural formula for $[Cu(H_2O)_6]^{2+}$ showing the shape of the complex. **1**

(iii) Explain why both these complexes are coloured and explain why changing the ligand from water to ammonia results in a change in colour. **3**

Cu+6(

MARKS | DO NOT WRITE IN THIS MARGIN

**5.** Barium carbonate decomposes on heating.

$BaCO_3(s) \rightarrow BaO(s) + CO_2(g) \quad \Delta H^0 = +266\,kJ\,mol^{-1}$

(a) Using the data from the table below, calculate the standard entropy change, $\Delta S^0$, in $J\,K^{-1}\,mol^{-1}$, for the reaction.

| *Substance* | *Standard entropy, $S^\circ / J\,K^{-1}\,mol^{-1}$* |
|---|---|
| $BaCO_3(s)$ | 112·0 |
| $BaO(s)$ | 72·1 |
| $CO_2(g)$ | 213·8 |

**2**

(b) Calculate the temperature at which the decomposition of barium carbonate just becomes feasible. **3**

MARKS | DO NOT WRITE IN THIS MARGIN

**6.** The bromate ion, $BrO_3^-$, is a useful oxidising agent.

(a) Calculate the oxidation number of bromine in the bromate ion. **1**

(b) The following table of results was obtained for the reaction between bromate ions and bromide ions under acidic conditions.

$BrO_3^-(aq) + 5Br^-(aq) + 6H^+(aq) \rightarrow 3Br_2(aq) + 3H_2O(l)$

| *Experiment* | *$[BrO_3^-]$/mol $l^{-1}$* | *$[Br^-]$/mol $l^{-1}$* | *$[H^+]$/mol $l^{-1}$* | *Initial rate/mol $l^{-1}$ $s^{-1}$* |
|---|---|---|---|---|
| 1 | 0·05 | 0·05 | 0·05 | $5{\cdot}0 \times 10^{-5}$ |
| 2 | 0·10 | 0·05 | 0·05 | $1{\cdot}0 \times 10^{-4}$ |
| 3 | 0·10 | 0·10 | 0·05 | $2{\cdot}0 \times 10^{-4}$ |
| 4 | 0·05 | 0·05 | 0·10 | $2{\cdot}0 \times 10^{-4}$ |

(i) The order of reaction with respect to $BrO_3^-$ is 1st order. Deduce the order of reaction with respect to $Br^-$ and $H^+$. **2**

(ii) Write the rate equation for the reaction. **1**

(iii) Calculate the rate constant for this reaction giving the appropriate units. **2**

MARKS DO NOT WRITE IN THIS MARGIN

**7.** The electronic spectra of molecules can be described in terms of the wavelength of maximum absorbance, $\lambda_{max}$.

The table below shows a number of compounds with their corresponding $\lambda_{max}$ values.

| *Compound* | $\lambda_{max}$/ nm |
|---|---|
| 1. | 217 |
| 2. | 227 |
| 3. | 263 |
| 4. | 352 |

(a) Compound 1 is buta-1,3-diene.

Name compound 2. **1**

(b) Draw the most likely structure for the compound with $\lambda_{max}$ = 291 nm. **1**

(c) The compounds shown have a system of alternating single and double bonds.

What word is used to describe this type of system? **1**

(d) From the information shown in the table draw a conclusion relating the energy difference between the HOMO and the LUMO as the number of alternating single and double bonds increases. **1**

MARKS DO NOT WRITE IN THIS MARGIN

**7. (continued)**

(e) β-carotene, $\lambda_{max}$ = 452 nm gives the orange colour to carrots and has the structure

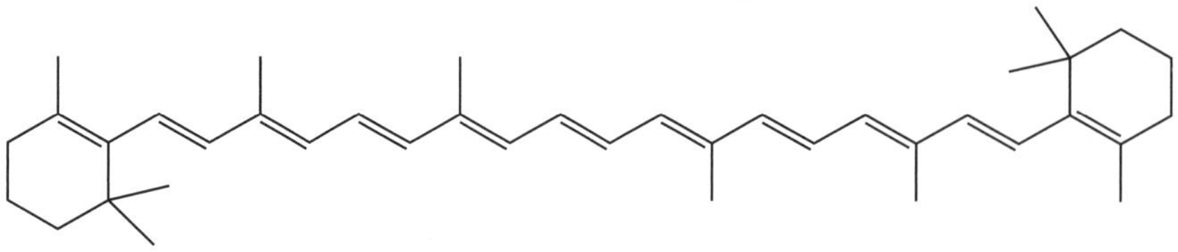

whereas α-carotene, $\lambda_{max}$ = 434 nm is found in oranges and has the structure

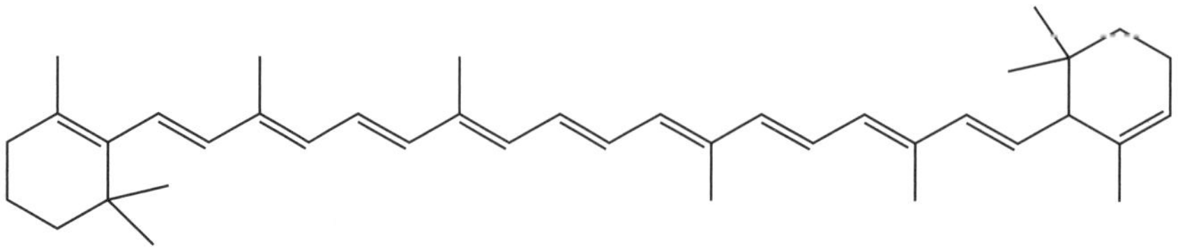

Explain why there is a difference in the $\lambda_{max}$ values for these two structures. **1**

(f) The pink colour of cooked salmon and lobster is due to astaxanthin which has the structure

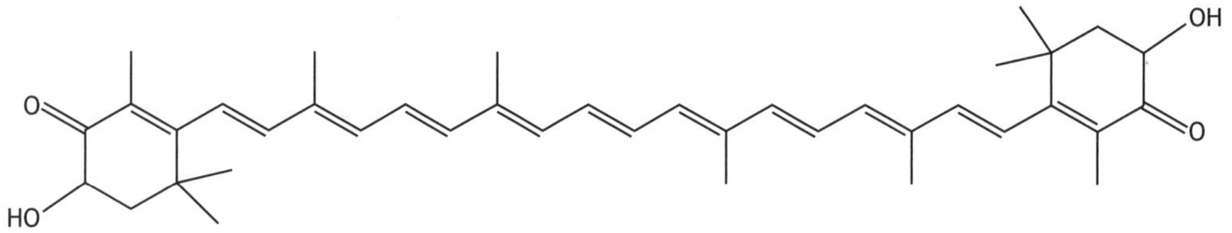

O HO OH O

This molecule is optically active.

Draw part of the molecule and circle one of the asymmetric carbon atoms responsible for this optical activity. **1**

MARKS DO NOT WRITE IN THIS MARGIN

**8.** 2-Methoxy-2-methylpropane is a compound added to unleaded petrol as a 'knock inhibitor'. It can be synthesised by the reaction of methoxide ions with 2-chloro-2-methylpropane.

$$(H_3C)_3C{-}Cl + H_3C{-}O^- \longrightarrow (H_3C)_3C{-}O{-}CH_3$$

2-chloro-2-methylpropane    methoxide ion    2-methoxy-2-methylpropane

(a) To which class of organic compounds does 2-methoxy-2-methylpropane belong? **1**

(b) How can the methoxide ion be prepared from methanol? **1**

(c) The preparation of the 2-methoxy-2-methylpropane proceeds by a $S_N1$ mechanism.

(i) Clearly showing the electron shifts, outline the step(s) involved. **3**

(ii) Suggest why this reaction is more likely to proceed by a $S_N1$ mechanism rather than a $S_N2$. **1**

(d) 2-Methoxybutane is an isomer of 2-methoxy-2-methylpropane. When 2-methoxybutane is created, a racemic mixture is formed.

(i) What is a racemic mixture? **1**

(ii) What would happen to polarised light if it was passed through a racemic mixture? **1**

MARKS DO NOT WRITE IN THIS MARGIN

**9.** A laboratory technician prepared a solution of potassium iodate, $KIO_3(aq)$, for use in a titration.

(a) Potassium iodate is an example of a primary standard. State one of the characteristics of a primary standard. **1**

(b) Describe the steps the technician should take to prepare a standard solution of potassium iodate. **2**

(c) Describe how $100\,cm^3$ of $0{\cdot}10\,mol\,l^{-1}$ potassium iodate solution would be prepared from a standard $1{\cdot}00\,mol\,l^{-1}$ potassium carbonate solution. **2**

MARKS | DO NOT WRITE IN THIS MARGIN

**10.** Methyl propanoate can be made by reacting methanol with propanoic acid.

$CH_3OH + CH_3CH_2COOH \leftrightharpoons CH_3OOCCH_2CH_3 + H_2O$

(a) State the type of reaction occurring when methyl propanoate is formed from methanol and propanoic acid. **1**

(b) Propanoic acid can be formed from the acid hydrolysis of propane nitrile. Draw a structural formula for propane nitrile. **1**

(c) An $^1H$ NMR spectrum of methyl propanoate is shown below. On the structure, label the H atoms corresponding to Hc. **1**

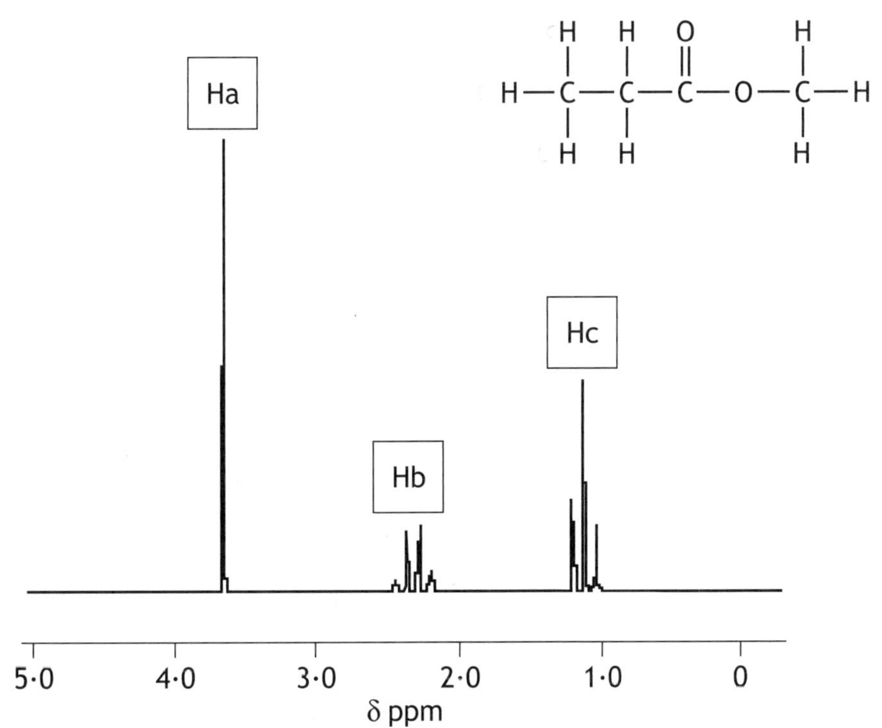

MARKS | DO NOT WRITE IN THIS MARGIN

**10. (continued)**

(d) A student suggested that the conversion of methanol and propanoic acid into methyl propanoate could be followed using analytical techniques such as chromatography and infrared spectroscopy allowing a chemist to know when the product ester had formed. Using your knowledge of chemistry, explain how these techniques could be used in this way. **3**

MARKS | DO NOT WRITE IN THIS MARGIN

**11.** Sulfa drugs are compounds with antibiotic properties. Sulfa drugs can be prepared from a solid compound called sulfanilamide.

Sulfanilamide is prepared in a six-stage synthesis. The equation for the final step in the synthesis is shown.

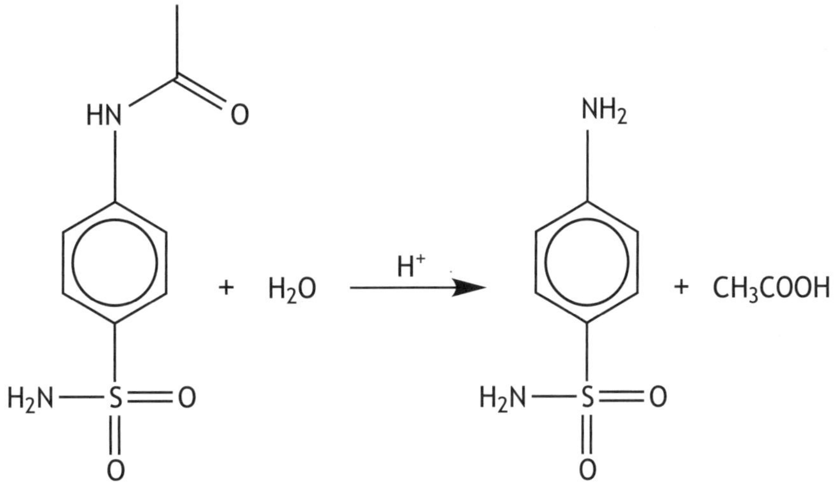

HN–C(=O)CH$_3$ / benzene ring / $H_2N$–S(=O)$_2$ + $H_2O$ $\xrightarrow{H^+}$ $NH_2$ / benzene ring / $H_2N$–S(=O)$_2$ + $CH_3COOH$

$C_8H_{10}N_2SO_3$ 4-acetamidobenzenesulfonamide

$C_6H_8N_2SO_2$ sulfanilamide

(a) Converting the 4-acetamidobenzesulfonamide into sulfanilamide involves gentle heating of the 4-acetamidobenzesulfonamide in a solution of water and hydrochloric acid in a round bottom flask for at least 30 minutes.

Name two pieces of equipment necessary to carry out this reaction successfully. **2**

(b) The sulfanilamide is separated from the reaction mixture and recrystallised from boiling water.

(i) Why is recystallisation necessary? **1**

(ii) Describe the solubilty of pure sulfanilamide in cold and hot water. **1**

(iii) Simple filtration to isolate the pure sulfanilamide is very slow.

How could the filtration be speeded up? **1**

MARKS DO NOT WRITE IN THIS MARGIN

**11. (continued)**

(c) Calculate the percentage yield of sulfanilamide if 4·282 g of 4-acetamidobenzenesulfonamide produced 2·237 g of sulfanilamide. **3**

(d) Describe how a mixed melting point experiment would be carried out and the result used to confirm that the product was pure. **2**

(e) Suggest another analytical technique which could be used to indicate whether the final sample is pure. **1**

**[END OF MODEL PAPER]**

## ADDITIONAL SPACE FOR ANSWERS AND ROUGH WORK

**ADDITIONAL SPACE FOR ANSWERS AND ROUGH WORK**

ADVANCED HIGHER

# Model Paper 2

Whilst this Model Paper has been specially commissioned by Hodder Gibson for use as practice for the Advanced Higher (for Curriculum for Excellence) exams, the key reference document remains the SQA Specimen Paper 2015 and SQA Past Paper 2016.

# Chemistry
# Section 1 — Questions

Duration — 2 hours 30 minutes

Instructions for the completion of Section 1 are given on *Page two* of your question and answer booklet.

Record your answers on the answer grid on *Page three* of your question and answer booklet.

Reference may be made to the Chemistry Higher and Advanced Higher Data Booklet.

Before leaving the examination room you must give your question and answer booklet to the Invigilator; if you do not you may lose all the marks for this paper.

**SECTION 1 — 30 marks**

**Attempt ALL questions**

1. Element X can be ionised to form $X^{2+}$ which has the electronic configuration $1s^2 2s^2 2p^6$.

   In which block of the periodic table would X be found?

   A s

   B p

   C d

   D f

2. A student observed a line emission spectrum by using a spectroscope to view the yellow colour emitted by a sodium vapour lamp. Which statement best describes the line emission spectrum for sodium?

   The line emission spectrum results from

   A the emission of photons as electrons drop to lower energy levels

   B the emission of electrons as they drop from high to low energy levels

   C electrons absorbing energy to move from lower energy d orbitals to higher energy d orbitals

   D electrons absorbing energy to move from the LUMO to the HOMO.

3. Which of the following statements for the first row transition metals is always true?

   A 2+ ions are formed by removing electrons from 3d orbitals.

   B The 3d orbitals remain degenerate when ligands bond to the metal.

   C All electronic configurations obey the Aufbau principle.

   D They have at least one electron in the 4s orbital.

4. The orbital box notation shown below for $1s^2$ does not conform to Pauli's exclusion principle.

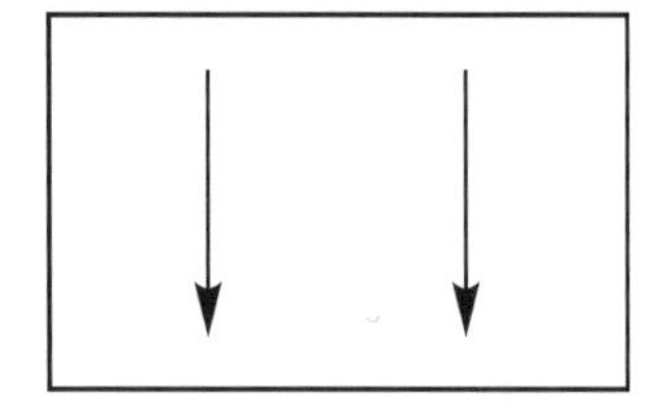

   Which quantum number should be changed to give the correct $1s^2$ orbital box notation?

   A Principal quantum number

   B Angular momentum quantum number

   C Magnetic quantum number

   D Spin quantum number

**5.** Which line in the graph represents the trend in successive ionisation energies of a Group 2 element?

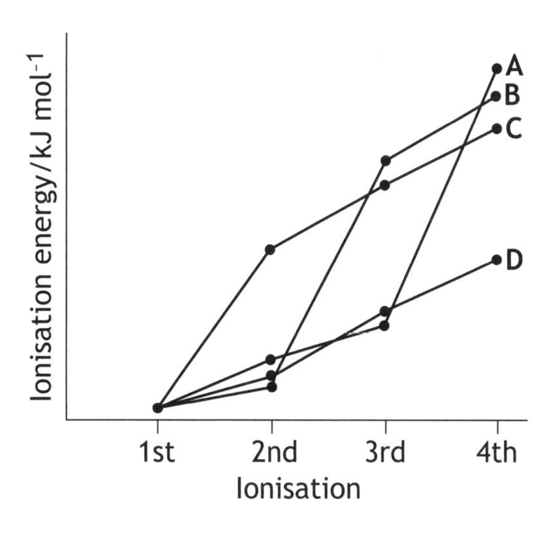


**6.** A dative covalent bond is present in

A $NH_3$

B $NH_4^+$

C $CH_3NH_2$

D $CH_3NH^-$.

**7.** For the following equilibrium

$N_2(g) + 3H_2(g) \rightleftharpoons 2NH_3(g)$ $\Delta H = -52 kJ\,mol^{-1}$

the value of K could be decreased by

A using an iron catalyst

B increasing the temperature

C increasing the pressure

D carrying out the reaction in an open system.

**8.** Equal concentrations of the following salt solutions were prepared.

Which of the following would have the highest pH value?

A Potassium methanoate

B Potassium chloride

C Potassium ethanoate

D Potassium benzoate

**9.** The pH of a buffer solution created from mixing equal volumes of 0·01 mol $l^{-1}$ methanoic acid with 0·05 mol $l^{-1}$ sodium methanoate would be

A 3·75

B 3·05

C 4·45

D 2·88.

**10.** Which of the following will lead to a decrease in entropy?

A Water evaporating

B Combustion of methanol

C Dehydration of ethanol

D Hydration of ethene

**11.** $CaCO_3(s) \rightarrow CaO(s) + CO_2(g)$

$\Delta H^0 = 178\ kJ\ mol^{-1}$ $\Delta S^0 = 161\ J\ K^{-1}\ mol^{-1}$

This reaction is thermodynamically feasible

A at any temperature

B at all temperatures below 904 K

C at all temperatures above 904 K

D only at temperatures above 1106 K.

**12.** NO(g) and $Cl_2(g)$ react according to the following equation:

$2NO(g) + Cl_2(g) \rightarrow 2NOCl(g)$

Which of the following statements is false if the reaction is 1$^{st}$ order with respect to NO(g) and 1$^{st}$ order with respect to $Cl_2(g)$?

A The overall order of this reaction is 2.

B The rate equation is rate = k[NO][$Cl_2$].

C The units of k would be mol $l^{-1}\ s^{-1}$.

D Doubling the concentration of NO would double the reaction rate.

**13.** A kinetics study of the reaction A + B $\rightarrow$ C gave the results shown in the table below.

| *Initial [A]/mol $l^{-1}$* | *Initial [B]/mol $l^{-1}$* | *Time to form a fixed concentration of C/s* |
|---|---|---|
| 0·2 | 0·2 | 24 |
| 0·2 | 0·4 | 12 |
| 0·4 | 0·2 | 24 |

The rate equation for the reaction is

A rate = $k[A][B]^2$

B rate = $k[B]^2$

C rate = $k[A][B]$

D rate = $k[B]$.

**14.** Which of the following could be applied to the bonding in propane?

A Sigma and pi bonds

B $sp^3$ and $sp^2$ hybrid orbitals

C $sp^3$ hybrid orbitals and sigma bonds

D $sp^3$ hybrid orbitals and p orbital overlap to form pi bonds

**15.** Which of the following molecules does not contain a conjugated system?

| | |
|---|---|
| **A** | 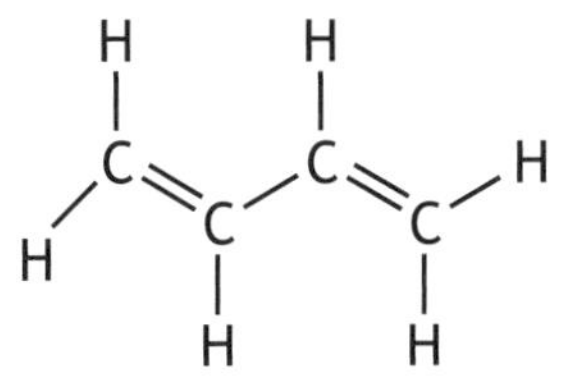 |
| **B** | 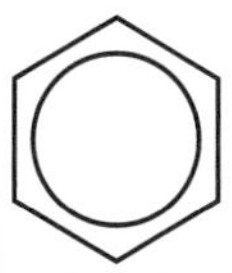 |
| **C** | $CH_2{=}CH_2{-}C{\equiv}N$ |
| **D** | 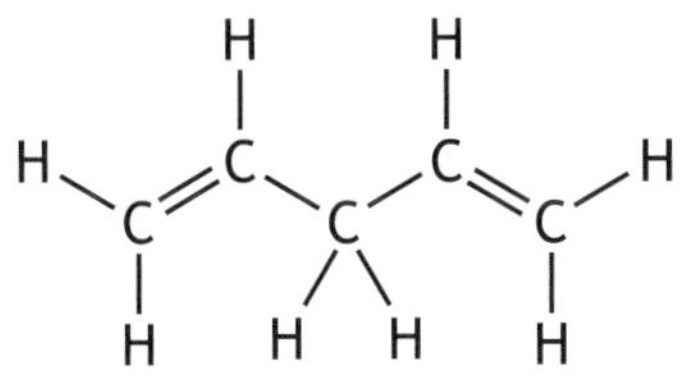 |

**16.**

OH

is the skeletal formula for

A pentan-2-ol

B 2-methylbutan-4-ol

C 2-methylbutan-2-ol

D 3-methylbutan-2-ol.

**17.** Which of the following is an example of a tertiary haloalkane?

A $CH_3Cl$

B $(CH_3)_3CCl$

C $(CH_3)_2CHCl$

D $(CH_2Cl)_4C$

**18.** Which of the following statements could not be applied to a racemic mixture?

A It is composed of two enantiomers.

B It would cause a rotation of polarised light.

C It contains two compounds which are non-superimposable mirror images.

D It contains optical isomers.

**19.** Which of the following equations does **not** represent a nucleophilic substitution?

A $C_3H_7Br + KOH \xrightarrow{\text{ethanol}} C_3H_6 + KBr + H_2O$

B $C_3H_7Br + KCN \xrightarrow{\text{ethanol}} C_3H_7CN + KBr$

C $C_2H_5Cl + C_2H_5ONa \xrightarrow{\text{ethanol}} C_2H_5OC_2H_5 + NaCl$

D $C_2H_5Br + NaOH \xrightarrow{\text{water}} C_2H_5OH + NaBr$

**20.** Which of the following reactions does not involve heterolytic fission?

A $C_2H_6 + Br_2 \rightarrow C_2H_5Br + HBr$

B $C_2H_4 + Br_2 \rightarrow C_2H_4Br_2$

C $C_2H_4 + HBr \rightarrow C_2H_5Br$

D $CH_3OH + HCOOH \rightarrow CH_3OOCH + H_2O$

**21.** Which of the following reactions would not produce propanoic acid?

A Hydrolysis of propanenitrile

B Hydrolysis of methyl propanoate

C Oxidation of propan-2-ol

D Oxidation of propanal

**22.** Naphthalene and anthracene are examples of polycyclic aromatic hydrocarbons.

Naphthalene:

structural formula

molecular formula $C_{10}H_8$

Anthracene:

structural formula

The molecular formula of anthracene is

A $C_{14}H_{10}$

B $C_{12}H_{10}$

C $C_{14}H_{12}$

D $C_{12}H_{12}$.

**23.** In the reaction between benzene and nitric acid in the presence of concentrated sulfuric acid

$C_6H_6 + HNO_3 \longrightarrow C_6H_5NO_2 + H_2O$

A the benzene molecule acts as an electrophile

B the ion, $NO_3^-$, acts as a nucleophile

C the ion, $NO_2^+$, acts as an electrophile

D the $HNO_3$ is oxidised.

**24.** Comparing methyl amine, $CH_3NH_2$, with butyl amine, $C_4H_9NH_2$, methyl amine will have a

A higher volatility and a higher solubility in water

B higher volatility and a lower solubility in water

C lower volatility and a higher solubility in water

D lower volatility and a lower solubility in water.

**25.** Elemental analysis of an organic compound showed it contained 70·6% carbon, 23·5% oxygen and 5·9% hydrogen by mass.

The structural formula of the compound could be

A $CH_2CHCH_2COOH$

B (benzene ring with two adjacent –OH groups)
OH
OH

C
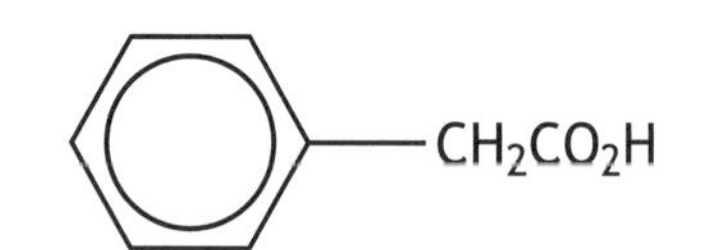


D (benzene ring)—$CH_2CHO$

**26.** An organic compound with empirical formula, $C_2H_4O$, has major peaks at 1715 $cm^{-1}$ and 3300 $cm^{-1}$ in its infrared spectrum.

The structural formula of the compound could be

A $CH_3CHO$

B $CH_3COOH$

C $CH_3COOCH_2CH_3$

D $CH_3CH_2CH_2COOH$.

**27.** Which of the following compounds could have produced the splitting pattern on an $^1H$ NMR spectrum as shown below?

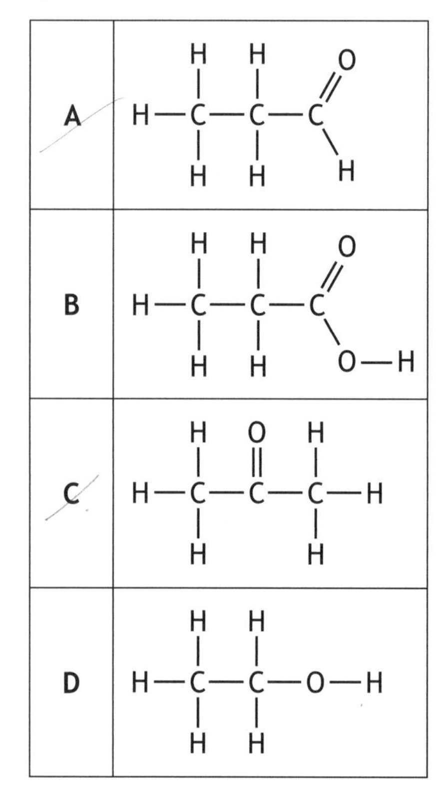


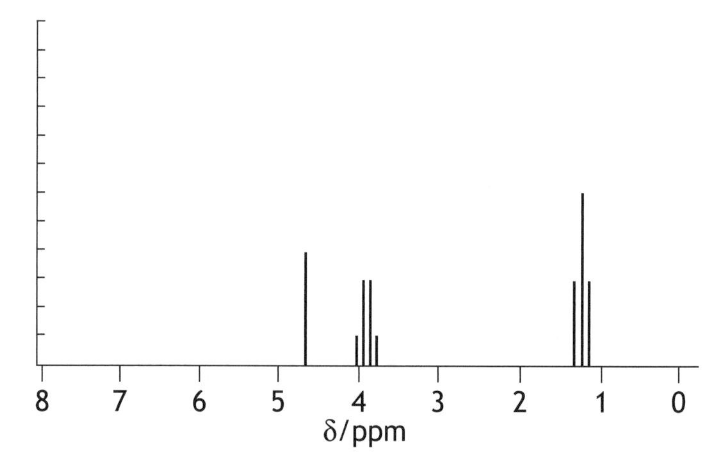

**28.** A drug containing a carboxyl group can bind to an amino group on a receptor site in three different ways.

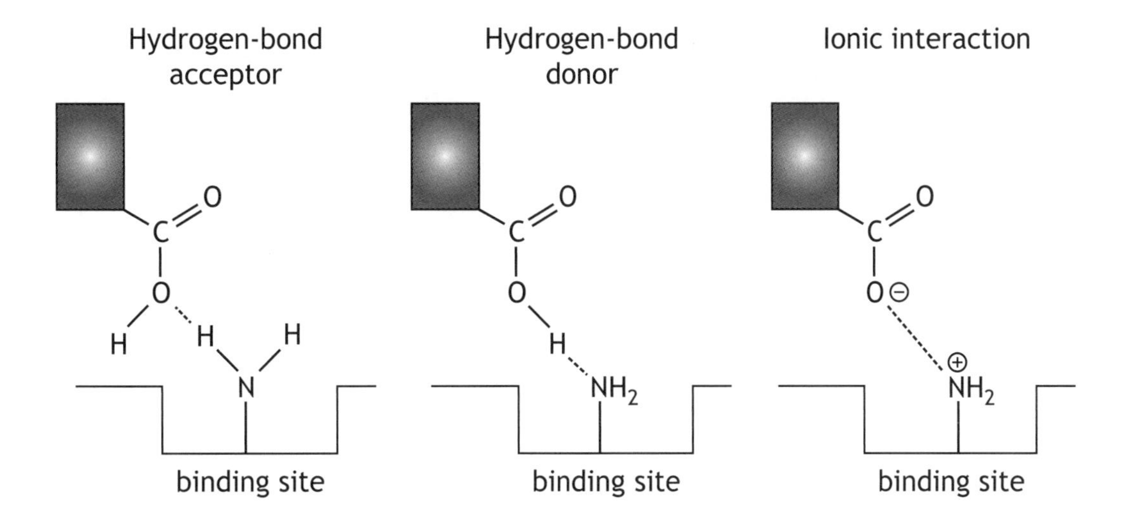


The drug with the following structure

could bind to the same site

A only by ionic interaction

B only as a hydrogen-bond donor

C only as a hydrogen-bond acceptor

D both as a hydrogen-bond donor and acceptor.

**29.** The stability of a covalent bond is related to its bond order, which can be defined as follows:

bond order = $\frac{1}{2}$ (number of bonding electrons – number of anti-bonding electrons)

The molecular orbital diagram for oxygen is shown. The anti-bonding orbitals are denoted by *.

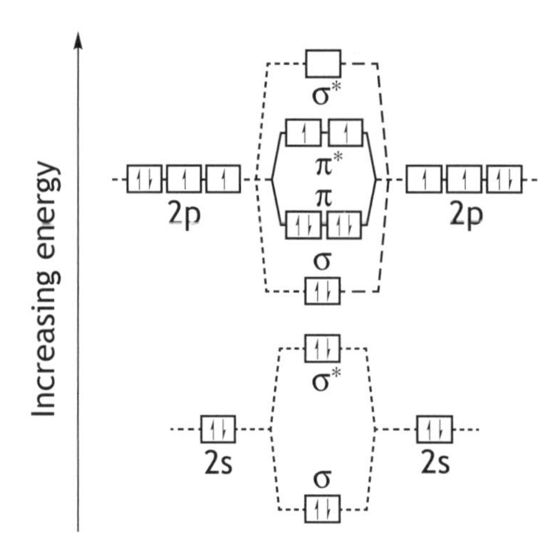


The bond order for a molecule of oxygen is

A 0

B 1

C 2

D 3.

**30.** Which one of the following is not suitable for the preparation of a primary standard in volumetric analysis?

A Anhydrous sodium carbonate

B Sodium hydroxide

C Oxalic acid

D Potassium iodate

**[END OF SECTION 1. NOW ATTEMPT THE QUESTIONS IN SECTION 2 OF YOUR QUESTION AND ANSWER BOOKLET.]**

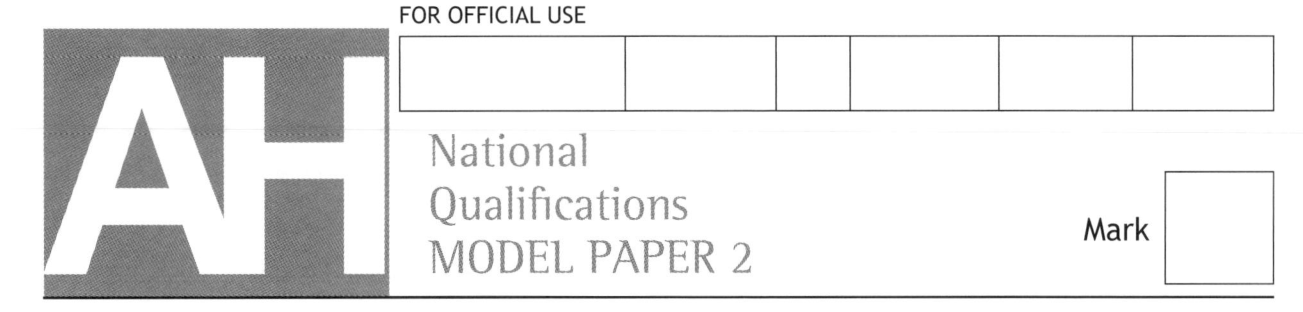

FOR OFFICIAL USE
AH
National
Qualifications
MODEL PAPER 2
Mark


# Chemistry
# Section 1 — Answer Grid and Section 2

Duration — 2 hours 30 minutes

**Fill in these boxes and read what is printed below.**

Full name of centre

Town

Forename(s)

Surname

Number of seat

Date of birth

Day Month Year

D D M M Y Y

Scottish candidate number

Reference may be made to the Chemistry Higher and Advanced Higher Data Booklet.

**Total marks — 100**

**SECTION 1 — 30 marks**

Attempt ALL questions.

Instructions for completion of Section 1 are given on *Page two*.

**SECTION 2 — 70 marks**

Attempt ALL questions

Write your answers clearly in the spaces provided in this booklet. Additional space for answers and rough work is provided at the end of this booklet. If you use this space you must clearly identify the question number you are attempting. Any rough work must be written in this booklet. You should score through your rough work when you have written your final copy.

Use **blue** or **black** ink.

Before leaving the examination room you must give this booklet to the Invigilator; if you do not, you may lose all the marks for this paper.

**SECTION 1— 30 marks**

The questions for Section 1 are contained in the question paper on *Page Four*.
Read these and record your answers on the answer grid on *Page three* opposite.
Do NOT use gel pens.

1. The answer to each question is **either** A, B, C or D. Decide what your answer is, then fill in the appropriate bubble (see sample question below).

2. There is **only one correct** answer to each question.

3. Any rough working should be done on the additional space for answers and rough work at the end of this booklet.

**Sample Question**

To show that the ink in a ball-pen consists of a mixture of dyes, the method of separation would be:

A fractional distillation

B chromatography

C fractional crystallisation

D filtration.

The correct answer is **B**—chromatography. The answer **B** bubble has been clearly filled in (see below).

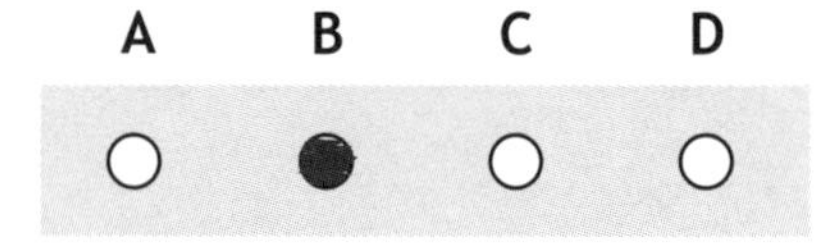


**Changing an answer**

If you decide to change your answer, cancel your first answer by putting a cross through it (see below) and fill in the answer you want. The answer below has been changed to **D**.

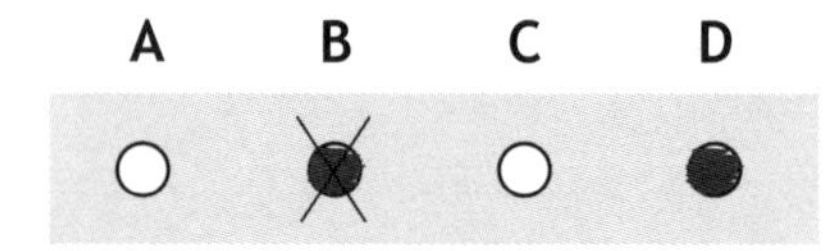


If you then decide to change back to an answer you have already scored out, put a tick (✓) to the **right** of the answer you want, as shown below:

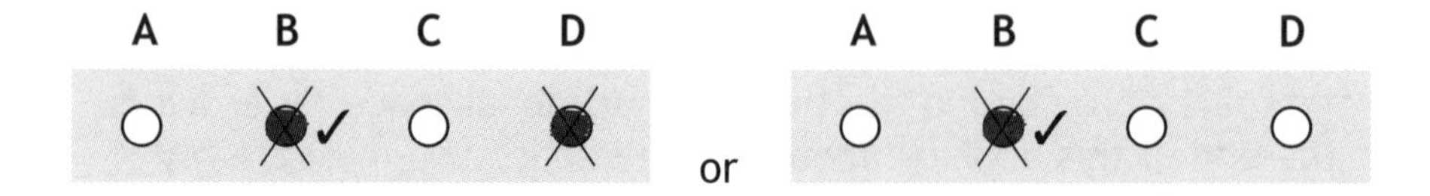

**SECTION 1 — Answer Grid**

| | A | B | C | D |
|---|---|---|---|---|
| 1 | ○ | ○ | ○ | ○ |
| 2 | ○ | ○ | ○ | ○ |
| 3 | ○ | ○ | ○ | ○ |
| 4 | ○ | ○ | ○ | ○ |
| 5 | ○ | ○ | ○ | ○ |
| 6 | ○ | ○ | ○ | ○ |
| 7 | ○ | ○ | ○ | ○ |
| 8 | ○ | ○ | ○ | ○ |
| 9 | ○ | ○ | ○ | ○ |
| 10 | ○ | ○ | ○ | ○ |
| 11 | ○ | ○ | ○ | ○ |
| 12 | ○ | ○ | ○ | ○ |
| 13 | ○ | ○ | ○ | ○ |
| 14 | ○ | ○ | ○ | ○ |
| 15 | ○ | ○ | ○ | ○ |
| 16 | ○ | ○ | ○ | ○ |
| 17 | ○ | ○ | ○ | ○ |
| 18 | ○ | ○ | ○ | ○ |
| 19 | ○ | ○ | ○ | ○ |
| 20 | ○ | ○ | ○ | ○ |
| 21 | ○ | ○ | ○ | ○ |
| 22 | ○ | ○ | ○ | ○ |
| 23 | ○ | ○ | ○ | ○ |
| 24 | ○ | ○ | ○ | ○ |
| 25 | ○ | ○ | ○ | ○ |
| 26 | ○ | ○ | ○ | ○ |
| 27 | ○ | ○ | ○ | ○ |
| 28 | ○ | ○ | ○ | ○ |
| 29 | ○ | ○ | ○ | ○ |
| 30 | ○ | ○ | ○ | ○ |

MARKS DO NOT WRITE IN THIS MARGIN

## SECTION 2 — 70 marks

**Attempt ALL questions**

**1.** Will-o'-the-wisp is a ghostly flickering flame that appears over marshes. According to folklore, it lured travellers from well-trodden paths into treacherous marshes. It is now known to be caused by the spontaneous combustion of a mixture of methane and phosphorus(III) hydride formed from rotting vegetation in the depths of marshes.

(a) Draw the shape of the phosphorus(III) hydride molecule. **1**

(b) Phosphorus(III) hydride contains the element phosphorus and the electronic configuration of a phosphorus atom in its ground state can be represented as:

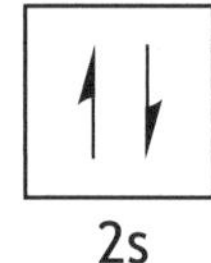
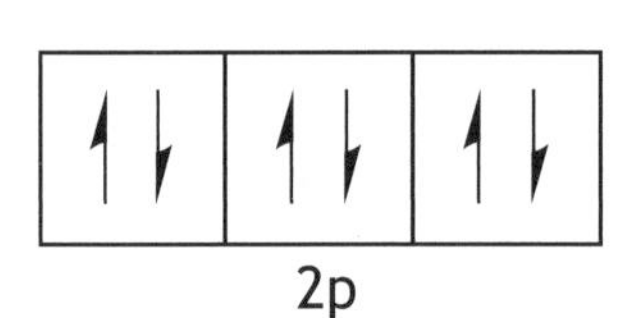
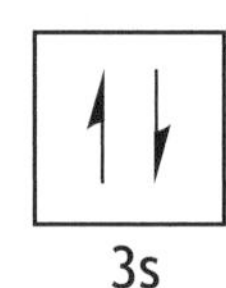
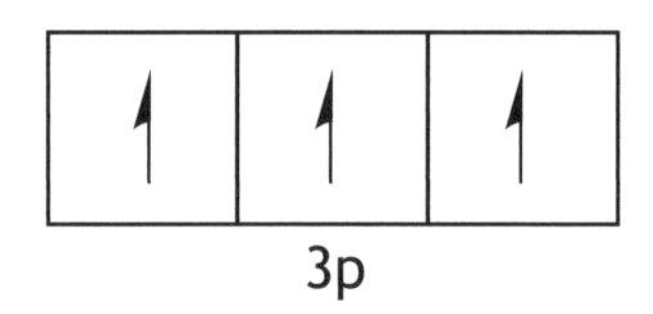

1s 2s 2p 3s 3p

(i) State the values of the four quantum numbers (***n***, ***l***, ***m*** and ***s***) for one of the electrons in the 3s orbital. **2**

(ii) In what respect is the 1s orbital similar to and how does it differ from the 2s orbital? **1**

MARKS | DO NOT WRITE IN THIS MARGIN

**1. (b) (continued)**

(iii) Explain how the above representation of the phosphorus atom conforms to Hund's rule. **1**

(iv) In an excited state of a phosphorus atom, one of its electrons has the quantum numbers $\boldsymbol{n}$ = 4 and $\boldsymbol{l}$ = 2.

In what type of orbital is this electron located? **1**

(c) The concentration of methane gas in a 2 kg sample of air was found to be 5000 ppm.

Calculate the number of moles of methane present. **2**

MARKS DO NOT WRITE IN THIS MARGIN

**2.** The nickel content of a compound can be determined by the three methods listed below.

- Volumetric
- Colorimetric
- Gravimetric

(a) In the volumetric method, a complexometric titration can be carried out.

1·33 g of hydrated nickel(II)sulfate, $NiSO_4{\cdot}6H_2O$, was weighed out accurately and made up to 100·0 cm$^3$ in a standard flask using distilled water.

25·0 cm$^3$ portions were titrated against 0·110 mol l$^{-1}$ EDTA solution using murexide indicator.

The results of the complexometric titrations are shown in the table below.

| | *Rough titration* | *1st titration* | *2nd titration* |
|---|---|---|---|
| **Initial burette reading/cm$^3$** | 0·0 | 12·0 | 23·7 |
| **Final burette reading/cm$^3$** | 12·0 | 23·7 | 35·5 |
| **Volume of EDTA added/cm$^3$** | 12·0 | 11·7 | 11·8 |

(i) From the experimental results, calculate the percentage of nickel, by mass, in the hydrated nickel(II) sulfate sample. **3**

MARKS | DO NOT WRITE IN THIS MARGIN

**2. (a) (continued)**

(ii) At the end point, the colour of the solution changes from yellow to violet.

Sketch the absorption curve for the violet solution below. **1**

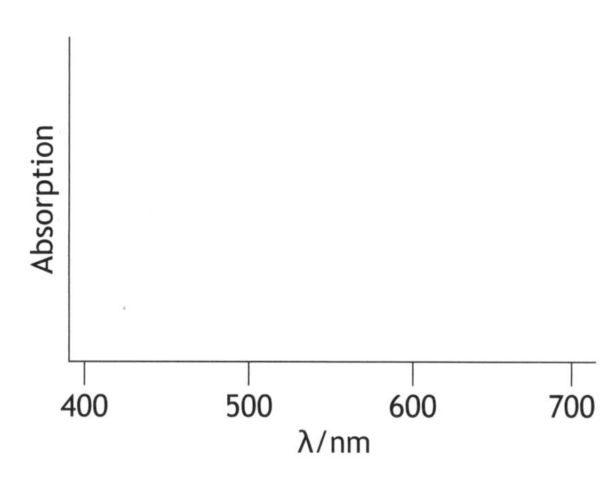


(b) A solution of nickel(II) sulfate is green in colour and so a pure sample of it can be used in a colorimetric method to determine the percentage of nickel in a solution whose concentration is not accurately known.

In the gravimetric method, a solution of nickel(II) sulfate reacts with dimethylglyoxime to produce a red solid. This can be filtered off, washed, dried and weighed to constant mass in order to determine the percentage of nickel by mass.

Considering all three methods involved, state one major factor in each method which could contribute to uncertainty in the final result. **3**

(c) A nickel solution was passed through a flame and the light emitted observed through a spectroscope. A line in the emission spectrum for nickel was observed at a wavelength of 508 nm.

Calculate the energy, in $kJ\,mol^{-1}$, associated with this wavelength. **2**

MARKS DO NOT WRITE IN THIS MARGIN

3. The huge variety of colour we observe on planet Earth can be traced to the movement of electrons in an atom or molecule.

Using your knowledge of chemistry, comment on how the colour we see is created by the movement of electrons. **3**

MARKS | DO NOT WRITE IN THIS MARGIN

**4.** Consider the following reaction.

$CS_2(g) + 4H_2(g) \rightleftharpoons CH_4(g) + 2H_2S(g)$

At 900 °C the equilibrium concentrations are:

$[CS_2] = 0{\cdot}012\ mol\ l^{-1}$ $[H_2] = 0{\cdot}0020\ mol\ l^{-1}$

$[H_2S] = 0{\cdot}00010\ mol\ l^{-1}$ $[CH_4] = 0{\cdot}0054\ mol\ l^{-1}$

(a) Write down the expression for the equilibrium constant, K, for this reaction. **1**

(b) Calculate the value of the equilibrium constant, K, at 900 °C. **1**

MARKS DO NOT WRITE IN THIS MARGIN

**5.** Carbonated soft drinks contain carbon dioxide gas which dissolves in water to produce carbonic acid:

$CO_2(aq) + H_2O(l) \rightleftharpoons H^+(aq) + HCO_3^-(aq)$ $\quad pK_a = 6{\cdot}35$

(a) Identify the conjugate base in this reaction. **1**

(b) Calculate the pH of a 0·01 mol l$^{-1}$ solution of carbonic acid. **2**

(c) A solution of carbonic acid was titrated with potassium hydroxide solution.

(i) Suggest a suitable indicator to determine the end point of this reaction. **1**

(ii) The solution formed at the end of the titration was added to carbonic acid.

Explain why the resultant solution can resist changes in pH even when small quantities of acid or base are added. **2**

MARKS | DO NOT WRITE IN THIS MARGIN

**6.** Burning magnesium continues to burn when placed in a gas jar of carbon dioxide according to the equation

$2Mg(s) + CO_2(g) \rightarrow 2MgO(s) + C(s)$

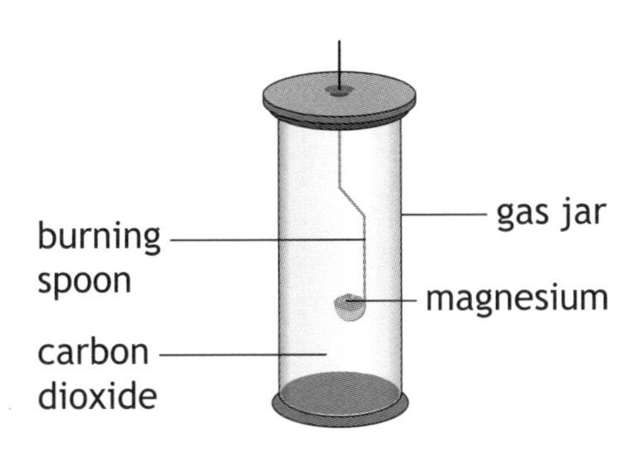


| *Substance* | *S°/J K$^{-1}$ mol$^{-1}$* |
|---|---|
| Mg(s) | 33·0 |
| $CO_2$(g) | 214·0 |
| MgO(s) | 27·0 |
| C(s) | 5·70 |

(a) Using the values from the table above, calculate ΔS° for the reaction. **1**

(b) Using the information below and your answer to (a), calculate ΔG° for the burning of magnesium in carbon dioxide at 298 K.

$Mg(s) + \frac{1}{2}O_2(g) \rightarrow MgO(s)$ $\quad \Delta H° = -493\ kJ\ mol^{-1}$

$C(s) + O_2(g) \rightarrow CO_2(g)$ $\quad \Delta H° = -394\ kJ\ mol^{-1}$ **3**

MARKS DO NOT WRITE IN THIS MARGIN

**7.** Consider the three reactions and their rate equations.

| | | |
|---|---|---|
| **Reaction 1** | $2N_2O_5 \rightarrow 4NO_2 + O_2$ | Rate = $k[N_2O_5]$ |
| **Reaction 2** | $2NO + Cl_2 \rightarrow 2NOCl$ | Rate = $k[NO]^2[Cl_2]$ |
| **Reaction 3** | $2NH_3 \rightarrow N_2 + 3H_2$ | Rate = $k[NH_3]^0$ |

(a) What is the overall order of Reaction 2? **1**

(b) The graph below was plotted using experimental results from one of the reactions.

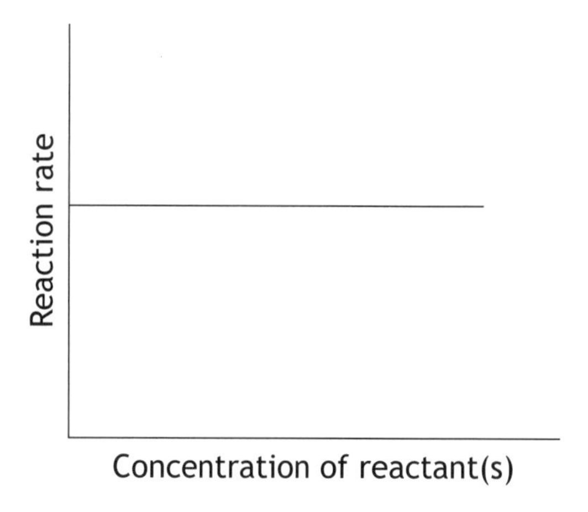


Explain which of the reactions would give this graph. **1**

(c) For Reaction 2, when the concentrations of NO and $Cl_2$ are both $0{\cdot}250\,mol\,l^{-1}$, the initial reaction rate is $1{\cdot}43 \times 10^{-6}\,mol\,l^{-1}\,s^{-1}$.

Use this information to calculate the rate constant, k, including the appropriate units. **2**

MARKS | DO NOT WRITE IN THIS MARGIN

**8.** The following procedure was used to determine the mass of aspirin present in an aspirin tablet.

**Step 1:** Crush an aspirin tablet and heat with excess sodium hydroxide solution.

$$\text{Aspirin} + 2NaOH \longrightarrow \text{(sodium salicylate: COONa, OH on benzene ring)} + CH_3COONa + H_2O(l)$$

Aspirin: benzene ring bearing COOH and O–C(=O)–$CH_3$

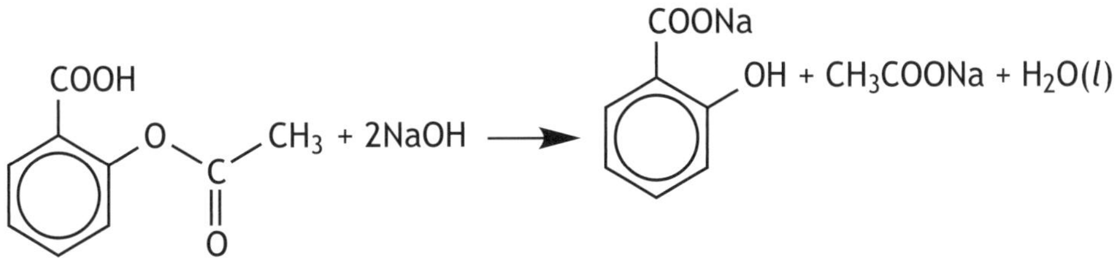

Aspirin
(1 mole = 180 g)

**Step 2:** Determine the excess sodium hydroxide by back titration using a standard solution of sulfuric acid.

$$2NaOH + H_2SO_4 \longrightarrow Na_2SO_4 + 2H_2O$$

An aspirin tablet was crushed and added to $50{\cdot}0\,cm^3$ of $1{\cdot}00\,mol\,l^{-1}$ sodium hydroxide solution. The solution was heated for 30 minutes. The cooled reaction mixture was then added to a $500\,cm^3$ standard flask, which was made up to the mark using deionised water. $25{\cdot}0\,cm^3$ samples were then titrated with $0{\cdot}10\,mol\,l^{-1}$ sulfuric acid until concordant results were obtained. The average titre was $12{\cdot}1\,cm^3$.

(a) Calculate the mass of aspirin in the tablet. **4**

(b) Suggest why a back titration technique has to be used to determine the mass of aspirin. **1**

MARKS | DO NOT WRITE IN THIS MARGIN

**8. (continued)**

(c) In a separate experiment, a student synthesised a sample of aspirin by reacting 2-hydroxybenzoic acid with excess ethanoic anhydride. They reported a 74% yield.

2-hydroxybenzoic acid (1 mole = 138 g) + ethanoic anhydride → aspirin (1 mole = 180 g) + $CH_3COOH$

(i) Give a reason why the percentage yield is not 100%. **1**

(ii) Calculate the mass of 2-hydroxybenzoic acid that should be reacted with excess ethanoic anhydride in order to produce 8 g of aspirin, assuming a 74% yield. **2**

(iii) The sample of aspirin obtained was analysed by thin layer chromatography and melting point analysis. Explain how both techniques can be used to help determine the purity of the aspirin. **2**

MARKS | DO NOT WRITE IN THIS MARGIN

**8. (continued)**

(d) Colorimetry can be used to determine the quantity of aspirin in a tablet. This is achieved by hydrolysing the aspirin to produce 2-hydroxybenzoic acid, which forms a coloured compound when reacted with iron(III) ions.

A method for the colorimetric determination of aspirin is shown below.

**Colorimetric determination of aspirin**

1. Weigh accurately about 0·4 g of aspirin into a 100 $cm^3$ conical flask, add 10 $cm^3$ of 1·0 $mol\,l^{-1}$ sodium hydroxide solution and warm the mixture gently at 50 °C for ten minutes.

2. Cool the solution and transfer quantitatively to a 500 $cm^3$ volumetric flask. Make up to the mark with deionised water. This is the stock solution which has a concentration equivalent to 0·8 $g\,l^{-1}$ of aspirin.

3. Use a burette to measure 10 $cm^3$ of stock solution into a 100 $cm^3$ volumetric flask. Make it up to volume with 0·02 $mol\,l^{-1}$ iron(III) chloride solution. This is standard solution A which has an equivalent aspirin concentration of 0·08 $g\,l^{-1}$. In a similar way make up further solutions and measure their absorbance using an appropriate filter on the colorimeter.

(i) Explain the meaning of the term "weigh accurately about 0·4 g". **1**

(ii) Show by calculation that the stock solution has a concentration equivalent to 0·8 $g\,l^{-1}$. **1**

(iii) Explain how the student would decide on "an appropriate filter" for this experiment. **1**

(iv) Suggest an improvement that could be made to step 3 to improve the accuracy of the results. **1**

MARKS DO NOT WRITE IN THIS MARGIN

**9.** Cyclohexanol can be converted into cyclohexene or cyclohexanone using different reagents as outlined below.

OH

①

cyclohexanol

cyclohexene

②

O

cyclohexanone

(a) Suggest a dehydrating agent that could be used to convert cyclohexanol into cyclohexene in reaction 1. **1**

(b) **Using your knowledge of chemistry**, comment on how it could be established that the product of reaction 2 is cyclohexanone. **3**

MARKS | DO NOT WRITE IN THIS MARGIN

**10.** Consider the following reaction sequence.

$$\mathbf{A} \xrightarrow{\text{HCN}} H_3C{-}\underset{\underset{\displaystyle CN}{|}}{\overset{\overset{\displaystyle OH}{|}}{C}}{-}H \xrightarrow{H_2O/H^+} H_3C{-}\underset{\underset{\displaystyle COOH}{|}}{\overset{\overset{\displaystyle OH}{|}}{C}}{-}H$$

**B** lactic acid

(a) HCN provides the $CN^-$ ion which attacks a positive carbon atom in reagent **A**.

Which term best describes reagent **A**? **1**

(b) Lactic acid in the form of lactate ions is dehydrogenated in the liver by the enzyme lactate dehydrogenase.

The diagram shows how one of the optical isomers of the lactate ion binds to an active site of lactate dehydrogenase.

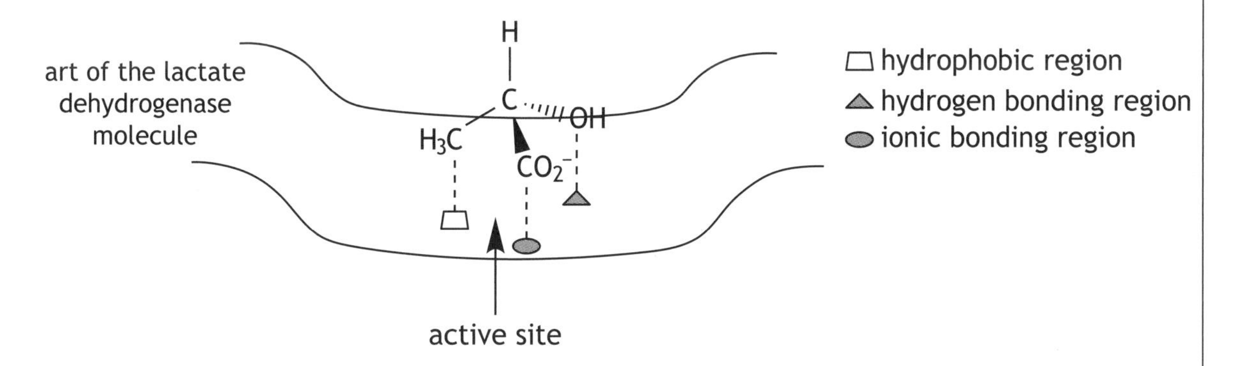


(i) Draw a structure for the other optical isomer of the lactate ion. **1**

(ii) Explain why this other optical isomer of the lactate ion cannot bind as efficiently to the active site of lactate dehydrogenase. **1**

MARKS DO NOT WRITE IN THIS MARGIN

**11.** A mixture of butan-1-ol and butan-2-ol can be synthesised from 1-bromobutane in a two-stage process.

$$CH_3CH_2CH_2CH_2Br \xrightarrow[\text{Stage 1}]{KOH/C_2H_5OH} CH_3CH_2CH{=}CH_2 \xrightarrow[\text{Stage 2}]{H_2O/H^+} \text{butan-1-ol} + \text{butan-2-ol}$$

(a) But-1-ene is produced after Stage 1. State the name of the other product formed at Stage 1. **1**

(b) The bonding in but-1-ene can be described in terms of $sp^2$ and $sp^3$ hybridisation and sigma and pi bonds.

(i) What is meant by $sp^2$ hybridisation? **1**

(ii) What is the difference in the way atomic orbitals overlap to form sigma and pi bonds? **1**

(c) Draw a structural formula for the major product of Stage 2. **1**

(d) 1-Bromobutane reacts with hydroxide ions to form butan-1-ol by an $S_N2$ mechanism.

Using structural formulae, draw the mechanism for $S_N2$ reaction of 1-bromobutane with hydroxide ions. **2**

MARKS | DO NOT WRITE IN THIS MARGIN

**12.** A compound **X** containing only carbon, hydrogen and oxygen was subjected to elemental analysis. Complete combustion of 1·76 g of **X** gave 3·52 g of carbon dioxide and 1·44 g of water. No other product was formed.

(a) (i) Calculate the masses of carbon and hydrogen in the original sample and hence deduce the mass of oxygen present. **2**

(ii) Show, by calculation, that the empirical formula of compound **X** is $C_2H_4O$. **1**

(b) Given that the relative molecular mass of compound **X** is 88, deduce its molecular formula. **1**

MARKS | DO NOT WRITE IN THIS MARGIN

**12. (continued)**

(c) A low-resolution proton NMR spectrum of compound **X** is shown.

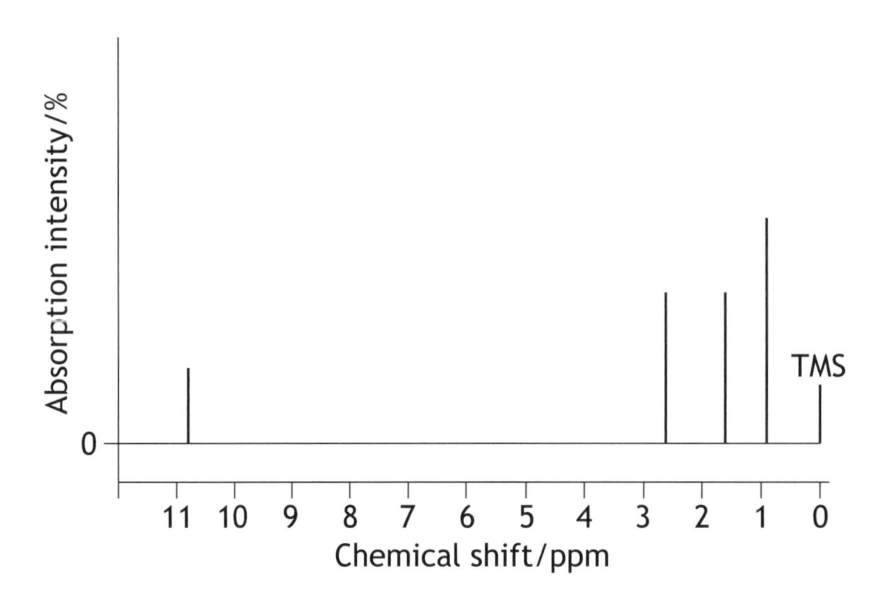


Analysis of this spectrum produced the data shown in the table below.

| *Chemical shift* | *Relative area under the peak* |
|---|---|
| 0·9 | 3 |
| 1·6 | 2 |
| 2·6 | 2 |
| 0·9<br>1·6<br>2·6<br>10·8 | 1 |

(i) Using information from the table and your answer to (b), draw a structural formula for compound **X** and give its systemic name. **2**

(ii) What is the function of 'TMS' (tetramethylsilane) in the proton NMR spectrum? **1**

**[END OF MODEL PAPER]**

**ADDITIONAL SPACE FOR ANSWERS AND ROUGH WORK**

**ADDITIONAL SPACE FOR ANSWERS AND ROUGH WORK**

ADVANCED HIGHER

# 2016

National Qualifications 2016

**X713/77/02**

# Chemistry
# Section 1 — Questions

WEDNESDAY, 18 MAY

9:00 AM – 11:30 AM

Instructions for the completion of Section 1 are given on *Page two* of your question and answer booklet X713/77/01.

Record your answers on the answer grid on *Page three* of your question and answer booklet.

Reference may be made to the Chemistry Higher and Advanced Higher Data Booklet.

Before leaving the examination room you must give your question and answer booklet to the Invigilator; if you do not, you may lose all the marks for this paper.

## SECTION 1 — 30 marks

**Attempt ALL questions**

1. Which of the following lists electromagnetic radiation bands in order of increasing wavelength?

   A X-ray, infrared, ultraviolet, radio

   B Infrared, ultraviolet, X-ray, gamma

   C Ultraviolet, visible, infrared, radio

   D Radio, infrared, visible, gamma

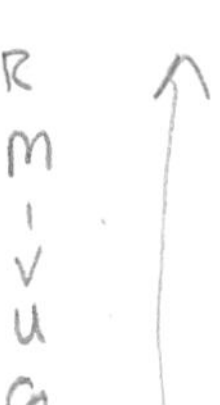

2. Which of the following states that electrons fill orbitals in order of increasing energy?

   A Hund's rule

   B The aufbau principle

   C The Pauli exclusion principle

   D The valence shell electron pair repulsion theory

3.

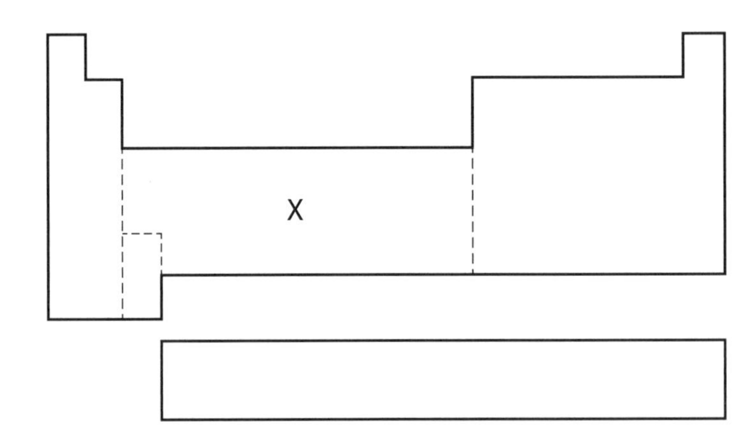


In the periodic table outlined above, one area is marked X. Moving across area X, from one element to the next, the extra electron usually occupies an orbital of type

   A s

   B p

   C d

   D f.

4. Which of the following molecules contains three atoms in a straight line?

   A $BF_3$

   B $CH_4$

   C $H_2O$

   D $SF_6$

**5.** The complex ion $[Cu(CN)_6]^{4-}$ is called

A hexacyanocopper(II)

B hexacyanocopper(IV)

C hexacyanocuprate(II)

D hexacyanocuprate(IV).

**6.** $HCN(aq) + H_2O(\ell) \rightleftharpoons H_3O^+(aq) + CN^-(aq)$

In the above equation HCN(aq) is acting as

A an acid

B a conjugate acid

C a base

D a conjugate base.

**7.** The use of an indicator is **not** appropriate in titrations involving

A hydrochloric acid solution and methylamine solution

B nitric acid solution and potassium hydroxide solution

C methanoic acid solution and ammonia solution

D propanoic acid solution and sodium hydroxide solution.

**8.** Which of the following can produce a buffer solution when added to aqueous $NH_4Cl$?

A Ammonia

B Ethanoic acid

C Potassium chloride

D Ammonium sulfate

**9.** Which of the following reactions **cannot** be described as an enthalpy of formation?

A $Si(s) + 4Cl(g) \rightarrow SiCl_4(\ell)$

B $Mg(s) + \frac{1}{2}O_2(g) \rightarrow MgO(s)$

C $C(s) + 2H_2(g) + \frac{1}{2}O_2(g) \rightarrow CH_3OH(\ell)$

D $2C(s) + 3H_2(g) \rightarrow C_2H_6(g)$

**[Turn over**

**10.** Which of the following is likely to have the lowest standard entropy at 100 °C?

A Neon

B Mercury

C Sulfur

D Phosphorus

**11.** For the reaction

$2A + 2B \rightarrow C$

the rate equation is

rate $= k[A][B]^2$.

Which of the following could be a possible mechanism for this reaction?

A $A + B \rightarrow X$ (fast)
$X + A + B \rightarrow C$ (slow)

B $A + 2B \rightarrow X$ (slow)
$X + A \rightarrow C$ (fast)

C $2A + B \rightarrow X$ (slow)
$X + B \rightarrow C$ (fast)

D $2A + B \rightarrow X$ (fast)
$X + B \rightarrow C$ (slow)

**12.** Which line in the table has the correct number and type of bonds in the structure shown?

| | *Number of σ-bonds* | *Number of π-bonds* |
|---|---|---|
| A | 2 | 18 |
| B | 4 | 16 |
| C | 16 | 4 |
| D | 18 | 2 |

**13.** 5-Methylhept-3-ene-2-one is an aroma molecule found in some types of tea.

Which of the following shows a structural formula for the *trans*-isomer of 5-methylhept-3-ene-2-one?

A

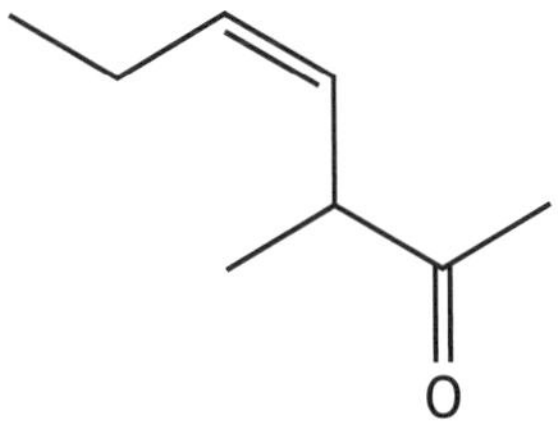

O

B

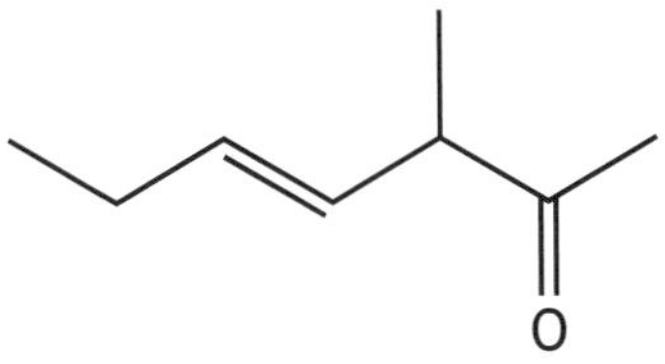

O

C

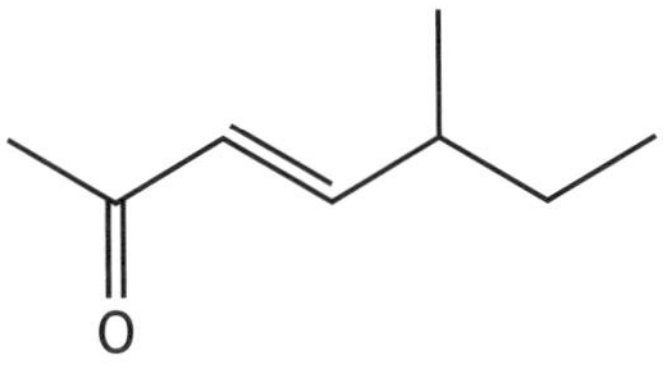

O

D

O

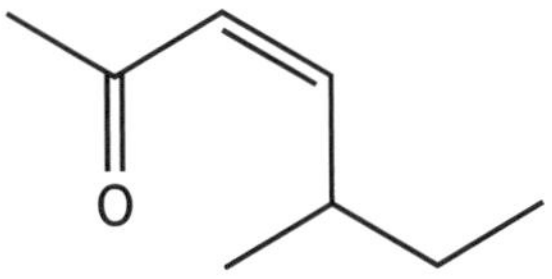

**14.** Which of the following does **not** exhibit hydrogen bonding between its molecules?

A Ethanol

B Ethylamine

C Ethanoic acid

D Ethoxyethane

**15.** In the homologous series of amines, an increase in chain length is accompanied by

| | *Volatility* | *Solubility in water* |
|---|---|---|
| A | increased | increased |
| B | decreased | decreased |
| C | increased | decreased |
| D | decreased | increased |



**[Turn over**

**16.** Which of the following will react together to produce 2-ethoxypropane?

A $CH_3CH_2OH$ and $CH_3CH_2COONa$

B $CH_3CH_2ONa$ and $CH_3CH_2CH_2Br$

C $CH_3CH(OH)CH_3$ and $CH_3COONa$

D $CH_3CH_2ONa$ and $CH_3CHBrCH_3$

**17.** Aldehydes can be converted into alcohols by the reaction shown

$$R'(H)C{=}O \xrightarrow[\text{(ii) } H^+]{\text{(i) } CH_3MgBr} H_3C{-}C(R')(H){-}OH$$

Which of the following aldehydes would produce a primary alcohol?

A Methanal

B Ethanal

C Propanal

D Butanal

**18.** $CH_3CHO + NH_2NH_2 \rightarrow CH_3CH = NNH_2 + H_2O$

This reaction is an example of

A hydration

B hydrolysis

C dehydration

D condensation.

**19.** When but-1-ene reacts with hydrogen chloride, 1-chlorobutane and 2-chlorobutane are formed. According to Markovnikov's rule

A there will be more 2-chlorobutane than 1-chlorobutane

B there will be more 1-chlorobutane than 2-chlorobutane

C there will be equal proportions of both products

D it is impossible to tell the relative proportion of each product.

**20.** When 2-bromobutane reacts with ethanolic potassium cyanide and the compound formed is hydrolysed with dilute acid, the final product is

A butanoic acid

B pentanoic acid

C 2-methylbutanoic acid

D 2-methylpentanoic acid.

**21.**

```
     H   H   H                    H   H   H
     |   |   |                    |   |   |
 H — C — C — C — H            H — C — C — C — H
     |   |   |                    |   |   |
     H   Cl  H        1           H   OH  H        2

     H       H                    H       H
     |       |                    |       |
 H — C — C — C — H            H — C — C = C
     |   ||  |                    |   |   |
     H   O   H        3           H   H   H        4
```

Which line in the table correctly identifies **W**, **X**, **Y** and **Z** in the reaction sequence?

$$W \xrightarrow{\text{reduction}} X \xrightarrow{\text{dehydration}} Y \xrightarrow{\text{addition}} Z$$

| | *W* | *X* | *Y* | *Z* |
|---|---|---|---|---|
| A | 1 | 4 | 2 | 3 |
| B | 3 | 2 | 1 | 4 |
| C | 3 | 2 | 4 | 1 |
| D | 4 | 1 | 2 | 3 |

**22.** Which of the following statements about benzene is **not** true?

A It is planar.

B It is susceptible to attack by electrophilic reagents.

C Its carbon to carbon bonds are equal in length.

D It is readily attacked by bromine.

**[Turn over**

**23.** $(CH_3)_3CBr + OH^- \rightarrow (CH_3)_3COH + Br^-$

The above reaction proceeds via an $S_N1$ mechanism.

What effect will doubling the concentration of hydroxide ions have on the reaction rate?

A It will have no effect.

B The reaction rate will halve.

C The reaction rate will double.

D The reaction rate will increase by a factor of four.

**24.**

```
     H   H   O
     |   |   ||
 H — C — C — C — O — H
     |   |
     H  (H)
```

Which of the following shows the splitting pattern for the circled **H** atom above, in a high resolution proton NMR spectrum?

A

B

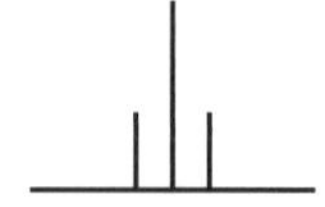

C

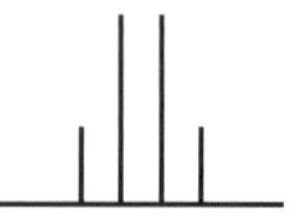

D

**25.**

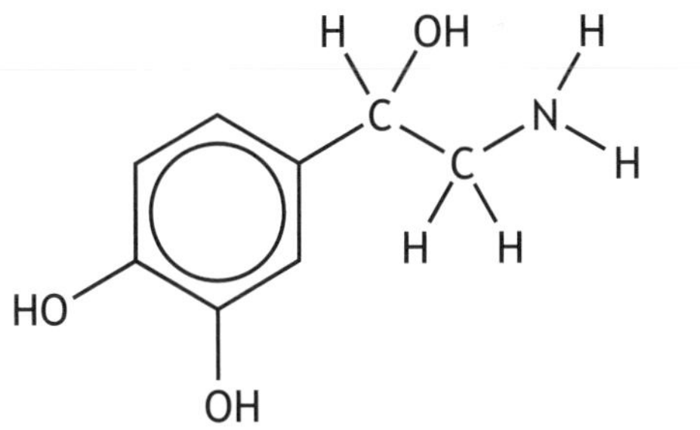

Noradrenaline

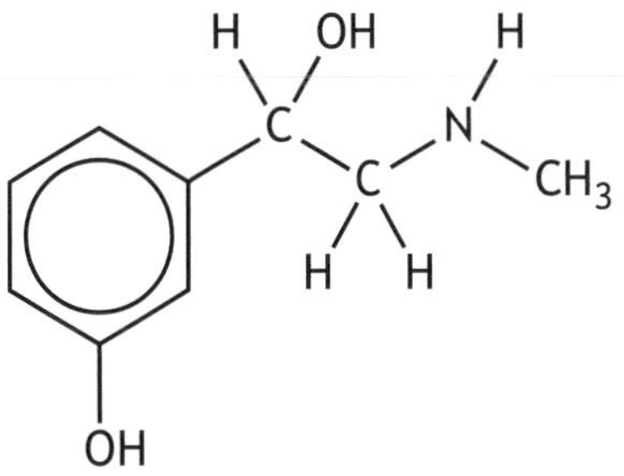

Phenylephrine

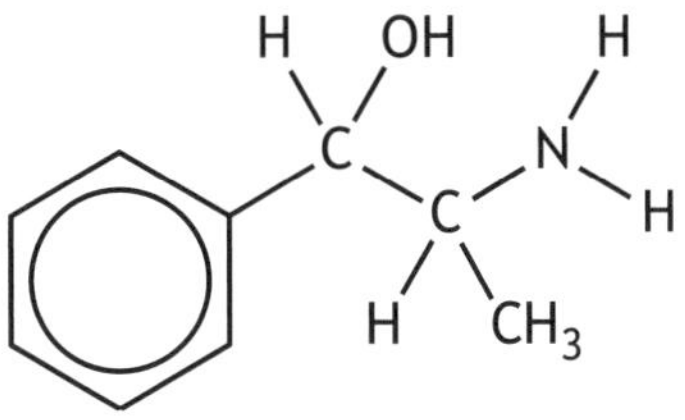

Amphetamine

Noradrenaline and phenylephrine stimulate receptors in the body resulting in increased blood pressure. Amphetamine has the same effect but works indirectly in the body by stimulating production of noradrenaline.

The structural fragment acting **directly** on the receptor is

A

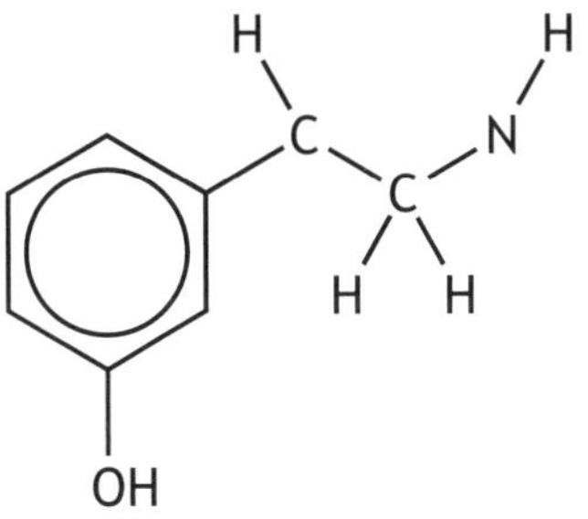

B

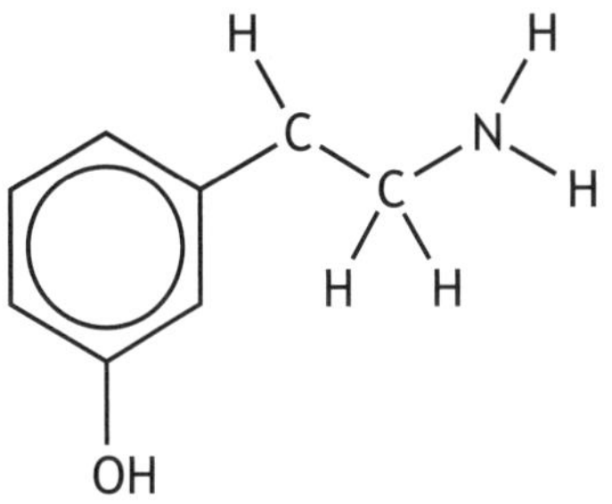

C

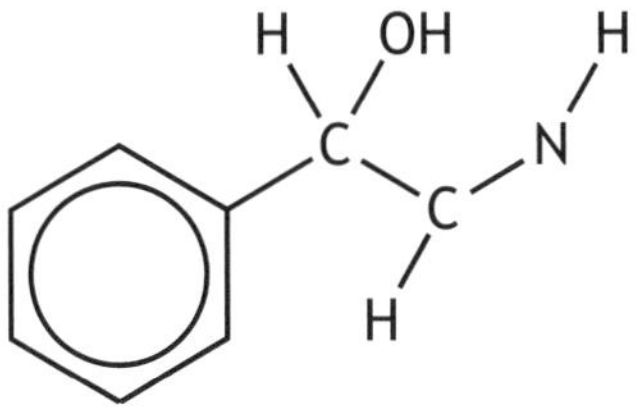

D

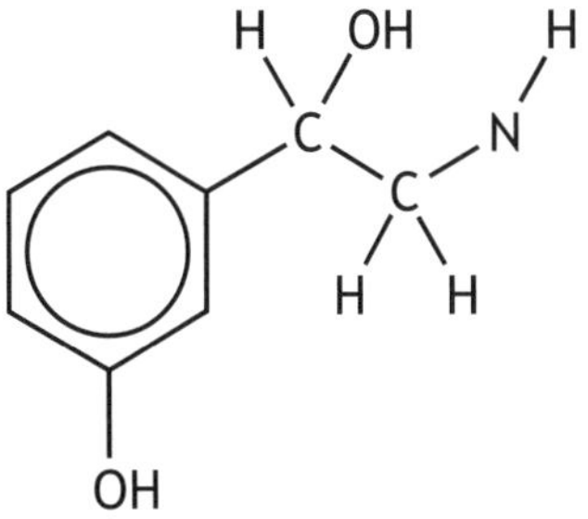

[Turn over

**26.** In a UK workplace, the maximum short-term exposure limit for carbon monoxide is 200 ppm in a 15 minute period.

If a person breathes in 134 g of air in a 15 minute period, what is the mass of carbon monoxide breathed in at the maximum short-term exposure limit?

A 1·49 mg

B 26·8 mg

C 1·49 g

D 26·8 g

**27.** Sodium hydroxide is unsuitable for use as a primary standard because it

A is corrosive

B is readily soluble in water

C is available in a high degree of purity

D readily absorbs water from the atmosphere.

**28.** What volume of 0·25 mol $l^{-1}$ calcium nitrate is required to make, by dilution with water, 500 $cm^3$ of a solution with a **nitrate** ion concentration of 0·1 mol $l^{-1}$?

A 50 $cm^3$

B 100 $cm^3$

C 200 $cm^3$

D 400 $cm^3$

**29.** 1·60 g of an anhydrous metal sulfate were dissolved in water. Addition of excess barium chloride solution resulted in the precipitation of 2·33 g of barium sulfate.

The original substance was

A copper(II) sulfate

B magnesium sulfate

C sodium sulfate

D calcium sulfate.

**30.** 0·020 moles of the salt $Pt(NH_3)_xCl_2$ required 20·0 $cm^3$ of 4·0 mol $l^{-1}$ nitric acid to react completely with the $NH_3$ ligands.

The value of x is

A 2

B 4

C 6

D 8.

**[END OF SECTION 1. NOW ATTEMPT THE QUESTIONS IN SECTION 2 OF YOUR QUESTION AND ANSWER BOOKLET]**

[BLANK PAGE]

DO NOT WRITE ON THIS PAGE

FOR OFFICIAL USE

National Qualifications 2016

Mark

# X713/77/01

# Chemistry Section 1 — Answer Grid and Section 2

WEDNESDAY, 18 MAY

9:00 AM – 11:30 AM

**Fill in these boxes and read what is printed below.**

Full name of centre

Town

Forename(s)

Surname

Number of seat

Date of birth

Day Month Year

Scottish candidate number

Reference may be made to the Chemistry Higher and Advanced Higher Data Booklet.

**Total marks — 100**

**SECTION 1 —30 marks**

Attempt ALL questions.

Instructions for the completion of Section 1 are given on *Page two.*

**SECTION 2 —70 marks**

Attempt ALL questions.

Write your answers clearly in the spaces provided in this booklet. Additional space for answers and rough work is provided at the end of this booklet. If you use this space you must clearly identify the question number you are attempting. Any rough work must be written in this booklet. You should score through your rough work when you have written your final copy.

Use **blue** or **black** ink.

Before leaving the examination room you must give this booklet to the Invigilator; if you do not, you may lose all the marks for this paper.

## SECTION 1 — 30 marks

The questions for Section 1 are contained in the question paper X713/77/02.

Read these and record your answers on the answer grid on *Page three* opposite.

Use **blue** or **black** ink. Do NOT use gel pens or pencil.

1. The answer to each question is **either** A, B, C or D. Decide what your answer is, then fill in the appropriate bubble (see sample question below).
2. There is **only one correct** answer to each question.
3. Any rough working should be done on the additional space for answers and rough work at the end of this booklet.

### Sample Question

To show that the ink in a ball-pen consists of a mixture of dyes, the method of separation would be:

A fractional distillation

B chromatography

C fractional crystallisation

D filtration.

The correct answer is **B** — chromatography. The answer **B** bubble has been clearly filled in (see below).

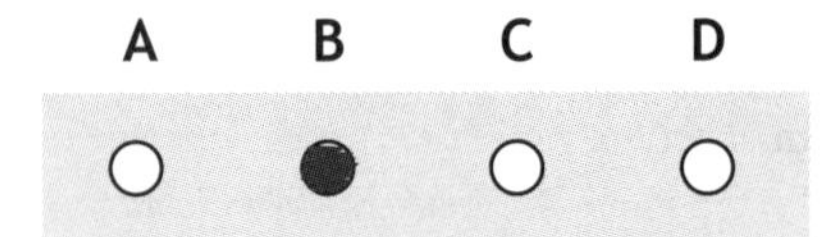


### Changing an answer

If you decide to change your answer, cancel your first answer by putting a cross through it (see below) and fill in the answer you want. The answer below has been changed to **D**.

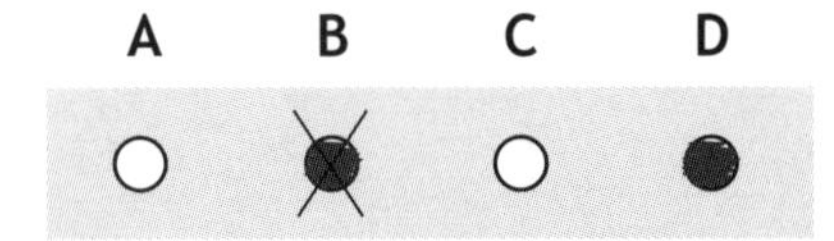


If you then decide to change back to an answer you have already scored out, put a tick (✓) to the **right** of the answer you want, as shown below:

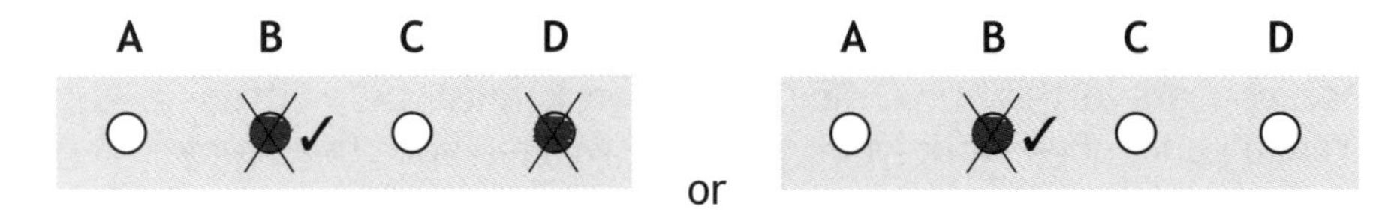

**SECTION 1 — Answer Grid**

| | A | B | C | D |
|---|---|---|---|---|
| 1 | ○ | ○ | ○ | ○ |
| 2 | ○ | ○ | ○ | ○ |
| 3 | ○ | ○ | ○ | ○ |
| 4 | ○ | ○ | ○ | ○ |
| 5 | ○ | ○ | ○ | ○ |
| 6 | ○ | ○ | ○ | ○ |
| 7 | ○ | ○ | ○ | ○ |
| 8 | ○ | ○ | ○ | ○ |
| 9 | ○ | ○ | ○ | ○ |
| 10 | ○ | ○ | ○ | ○ |
| 11 | ○ | ○ | ○ | ○ |
| 12 | ○ | ○ | ○ | ○ |
| 13 | ○ | ○ | ○ | ○ |
| 14 | ○ | ○ | ○ | ○ |
| 15 | ○ | ○ | ○ | ○ |

| | A | B | C | D |
|---|---|---|---|---|
| 16 | ○ | ○ | ○ | ○ |
| 17 | ○ | ○ | ○ | ○ |
| 18 | ○ | ○ | ○ | ○ |
| 19 | ○ | ○ | ○ | ○ |
| 20 | ○ | ○ | ○ | ○ |
| 21 | ○ | ○ | ○ | ○ |
| 22 | ○ | ○ | ○ | ○ |
| 23 | ○ | ○ | ○ | ○ |
| 24 | ○ | ○ | ○ | ○ |
| 25 | ○ | ○ | ○ | ○ |
| 26 | ○ | ○ | ○ | ○ |
| 27 | ○ | ○ | ○ | ○ |
| 28 | ○ | ○ | ○ | ○ |
| 29 | ○ | ○ | ○ | ○ |
| 30 | ○ | ○ | ○ | ○ |



**[Turn over**

[BLANK PAGE]

DO NOT WRITE ON THIS PAGE

[Turn over for next question

DO NOT WRITE ON THIS PAGE

MARKS | DO NOT WRITE IN THIS MARGIN

## SECTION 2 — 70 marks

**Attempt ALL questions**

1. Ethene can be hydrated to produce ethanol.

$$C_2H_4(g) + H_2O(\ell) \rightarrow C_2H_5OH(\ell)$$

| *Compound* | *Standard free energy of formation, $\Delta G^\circ$* (kJ mol$^{-1}$) | *Standard enthalpy of formation, $\Delta H^\circ{}_f$* (kJ mol$^{-1}$) |
|---|---|---|
| Ethene | 68 | 52 |
| Water | −237 | −286 |
| Ethanol | −175 | −278 |

(a) For the hydration of ethene, use the data in the table to calculate:

(i) the standard enthalpy change, $\Delta H^\circ$, in kJ mol$^{-1}$; **1**

(ii) the standard entropy change, $\Delta S^\circ$, in J K$^{-1}$ mol$^{-1}$. **3**

(b) Calculate the temperature, in K, at which this reaction just becomes feasible. **2**

MARKS | DO NOT WRITE IN THIS MARGIN

**2.** In the periodic table, period 2 is comprised of the elements lithium to neon.

The following table shows two of the quantum numbers for all ten electrons in a neon atom.

| *Electron* | *Principal quantum number, n* | *Angular momentum quantum number, l* |
|---|---|---|
| 1 | 1 | 0 |
| 2 | 1 | 0 |
| 3 | 2 | 0 |
| 4 | 2 | 0 |
| 5 | 2 | 1 |
| 6 | 2 | 1 |
| 7 | 2 | 1 |
| 8 | 2 | 1 |
| 9 | 2 | 1 |
| 10 | 2 | 1 |

(a) Write the electronic configuration for neon in terms of s and p orbitals. **1**

(b) The angular momentum quantum number, *l*, is related to the shape of an orbital.

Draw the shape of an orbital when *l* has a value of 1. **1**

(c) The magnetic quantum number, *m*, is related to the orientation of an orbital in space.

State the values of *m* for the orbital which contains the tenth electron. **1**

[BLANK PAGE]

DO NOT WRITE ON THIS PAGE

MARKS | DO NOT WRITE IN THIS MARGIN

**3.** Iron can form a variety of complexes with different ligands. Each complex has different properties.

(a) Some iron complex ions are paramagnetic. Paramagnetic substances are substances that are weakly attracted by a magnetic field.

Paramagnetism is caused by the presence of unpaired electrons.

In both $[Fe(H_2O)_6]^{2+}$ and $[Fe(CN)_6]^{4-}$, the $Fe^{2+}$ ion has six d-electrons, but only $[Fe(H_2O)_6]^{2+}$ is paramagnetic.

(i) Complete the d-orbital box diagram for the complex ion $[Fe(CN)_6]^{4-}$. **1**

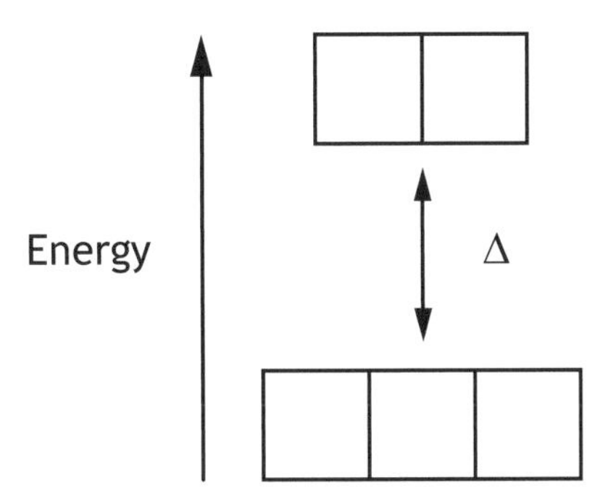


(An additional diagram, if required, can be found on *Page 28*)

(ii) The relative ability of a ligand to split the d-orbitals when forming a complex ion is given by the spectrochemical series.

The spectrochemical series for some ligands is shown below.

$$CN^- > NH_3 > H_2O$$

The $[Fe(H_2O)_6]^{2+}$ ion has unpaired electrons and is therefore paramagnetic.

Explain how unpaired electrons can arise in this complex ion. **2**

(iii) Explain why all of the complex ions formed by the $Fe^{3+}$ ion are paramagnetic. **1**

MARKS | DO NOT WRITE IN THIS MARGIN

**3. (continued)**

(b) Human blood is red due to the presence of haemoglobin bonded to oxygen. Other animals have different coloured blood due to the presence of different complex ions bonded to oxygen.

| Animal | Complex ion | Colour of blood |
|---|---|---|
| Human | Haemoglobin | RED |
| Spider | Oxyhaemocyanin | BLUE |
| Leech | Chlorocruorin | GREEN |

(i) State the co-ordination number of the $Fe^{2+}$ ion in haemoglobin. **1**

MARKS | DO NOT WRITE IN THIS MARGIN

**3. (b) (continued)**

(ii) Spiders' blood contains the oxyhaemocyanin complex ion. Oxyhaemocyanin contains copper ions.

Suggest an analytical technique that could be used to determine the presence of copper ions in spiders' blood. **1**

(iii) **Using your knowledge of chemistry**, comment on why these animals have different coloured blood. **3**

**[Turn over**

MARKS | DO NOT WRITE IN THIS MARGIN

**4.** As part of an Advanced Higher Chemistry project, a student determined the chloride ion concentration of seawater by two different methods.

*Volumetric method*
A sample of seawater was titrated with standard silver nitrate solution.

*Gravimetric method*
A sample of seawater was reacted with standard silver nitrate solution to form a precipitate. The precipitate was collected by filtration and weighed.

(a) For the volumetric method, a $0{\cdot}1\,mol\,l^{-1}$ standard solution of silver nitrate was prepared by following the instructions below.

1. Dry 5 g of silver nitrate for 2 hours at 100 °C and allow to cool.
2. Weigh accurately approximately 4·25 g of solid silver nitrate.
3. Use this sample to prepare $250\,cm^3$ of standard silver nitrate solution.

(i) State what is meant by "weigh accurately approximately" 4·25 g of solid silver nitrate. **1**

(ii) Outline how the student would have prepared the standard silver nitrate solution. **2**

(iii) Samples of the diluted seawater were titrated and the average titre was found to be $3{\cdot}9\,cm^3$.

Suggest an improvement the student could make to reduce the uncertainty in the titre value. **1**

MARKS DO NOT WRITE IN THIS MARGIN

4. (continued)

(b) For the gravimetric method, standard silver nitrate solution was added to a seawater sample to form a precipitate of silver chloride.

(i) Describe how the filtration should have been carried out to ensure a fast means of separating the precipitate from the reaction mixture. 1

(ii) After the precipitate was filtered, the filtrate was tested with a few drops of silver nitrate solution.

Suggest why the student tested the filtrate in this way. 1

(c) The student also planned to carry out an analysis of chloride ion concentration in fresh river water.

Explain why the volumetric method, rather than the gravimetric method, would be more appropriate for the analysis of chloride ion concentration in fresh river water. 1

[Turn over

MARKS | DO NOT WRITE IN THIS MARGIN

5. Mandelic acid, 2-hydroxy-2-phenylethanoic acid, is a component of skin care products.

mandelic acid

(a) Mandelic acid is a weak acid.

$$C_6H_5CH(OH)COOH(aq) + H_2O(\ell) \rightleftharpoons C_6H_5CH(OH)COO^-(aq) + H_3O^+(aq)$$

Write the expression for the dissociation constant, $K_a$, for mandelic acid. **1**

(b) A 100 $cm^3$ sample of skin care product contained 10·0 g of mandelic acid. The $K_a$ of mandelic acid is $1{\cdot}78 \times 10^{-4}$.

(i) Calculate the concentration of the mandelic acid, in $mol\,l^{-1}$, present in the skin care product. **2**

(ii) Using your answer to (b)(i), calculate the pH of a solution of mandelic acid of this concentration. **3**

[Turn over for next question

DO NOT WRITE ON THIS PAGE

MARKS | DO NOT WRITE IN THIS MARGIN

6. Chlorine is a versatile element which forms a wide range of compounds.

(a) One example of a compound containing chlorine is vanadium(IV) chloride. It reacts vigorously with water forming a blue solution.

The blue solution absorbs light of wavelength 610 nm.

Calculate the energy, in $kJ\,mol^{-1}$, associated with this wavelength. **2**

(b) Chlorine dioxide, $ClO_2$, is used in water sterilisation.

An experiment was carried out to determine the kinetics for the reaction between chlorine dioxide and hydroxide ions.

$$2ClO_2(aq) + 2OH^-(aq) \rightarrow ClO_2^-(aq) + ClO_3^-(aq) + H_2O(\ell)$$

Under certain conditions the following results were obtained.

| *[$ClO_2$]* ($mol\,l^{-1}$) | *[$OH^-$]* ($mol\,l^{-1}$) | *Initial rate* ($mol\,l^{-1}s^{-1}$) |
|---|---|---|
| $6{\cdot}00 \times 10^{-2}$ | $3{\cdot}00 \times 10^{-2}$ | $2{\cdot}48 \times 10^{-2}$ |
| $1{\cdot}20 \times 10^{-1}$ | $3{\cdot}00 \times 10^{-2}$ | $9{\cdot}92 \times 10^{-2}$ |
| $1{\cdot}20 \times 10^{-1}$ | $9{\cdot}00 \times 10^{-2}$ | $2{\cdot}98 \times 10^{-1}$ |

(i) Determine the order of reaction with respect to:

(A) $ClO_2$ **1**

(B) $OH^-$ **1**

MARKS DO NOT WRITE IN THIS MARGIN

**6. (b) (continued)**

(ii) Write the overall rate equation for the reaction. **1**

(iii) Calculate the value for the rate constant, $k$, including the appropriate units. **2**

MARKS | DO NOT WRITE IN THIS MARGIN

**7.** Aldehydes and ketones can exist in two forms, a keto form and an enol form.

For example, the aldehyde ethanal exists in equilibrium with its enol form, ethenol.

$$H_3C{-}CHO \rightleftharpoons H_2C{=}CH(OH) \qquad K = 3{\cdot}0 \times 10^{-7}$$

ethanal (keto form) — ethenol (enol form)

These two different molecules are known as tautomers.

(a) State which of the tautomers is the more abundant in this equilibrium. **1**

(b) 3-Methylpentan-2-one is optically active and exists in equilibrium with its enol tautomer.

(i) Circle the chiral centre on 3-methylpentan-2-one. **1**

$$H_3C{-}CH_2{-}C(H)(CH_3){-}C(=O){-}CH_3$$

(ii) Suggest why the optical activity of 3-methylpentan-2-one decreases over time. **1**

MARKS | DO NOT WRITE IN THIS MARGIN

**7. (b) (continued)**

(iii) Draw the skeletal formula for 3-methylpentan-2-one. **1**

(c) A possible mechanism for acid-catalysed enolisation is shown below, where R, R' and R" are alkyl groups.

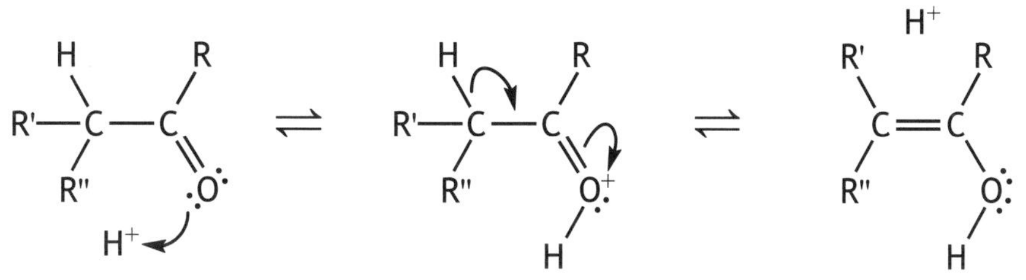

$$\mathrm{R'\!-\!C(H)(R'')\!-\!C(R)\!=\!\ddot{O}\!:} \;+\; \mathrm{H^+} \;\rightleftharpoons\; \mathrm{R'\!-\!C(H)(R'')\!-\!C(R)\!=\!O^+\!-\!H} \;\rightleftharpoons\; \mathrm{R'(R'')C\!=\!C(R)\!-\!\ddot{O}\!-\!H} \;+\; \mathrm{H^+}$$

Using structural formulae and curly arrow notation, show a possible mechanism for the acid-catalysed enolisation of 3-methylpentan-2-one. **3**

MARKS | DO NOT WRITE IN THIS MARGIN

**8.** Aspirin can be used as a starting material for the synthesis of the drug, salbutamol, which is used in the treatment of asthma. Salbutamol acts as an agonist by stimulating receptors in the lungs.

A possible synthetic route is shown.

$HO_2C$, $H_3COCO$ — Step ① $AlCl_3$ → $HO_2C$, HO, O

$CH_3OH/H^+$ Step ②

$H_3CO_2C$, HO, O, Br ← Step ③ $Br_2$ — $H_3CO_2C$, HO, O

Step ④

$H_3CO_2C$, HO, O, H, N, C, $CH_3$, $H_3C$, $CH_3$

Step ⑤

HO, HO, OH, H, N, C, $CH_3$, $H_3C$, $CH_3$

(a) State what is meant by the term agonist. **1**

(b) Step ① is known as a Fries rearrangement.

Suggest the role of $AlCl_3$ in this rearrangement. **1**

MARKS | DO NOT WRITE IN THIS MARGIN

**8. (continued)**

(c) Suggest a reaction condition required for Step ③. 1

(d) Identify the type of reaction taking place in Step ④. 1

(e) Step ⑤ involves several reactions.

Suggest a suitable reagent that could be used to convert the ketone carbonyl group to the hydroxyl group. 1

[Turn over

MARKS DO NOT WRITE IN THIS MARGIN

8. **(continued)**

(f) The purity of salbutamol can be determined using a variety of analytical techniques.

**Using your knowledge of chemistry**, discuss how analytical techniques could be used to determine the purity of salbutamol. 3

MARKS | DO NOT WRITE IN THIS MARGIN

**9.** Parabens are used as preservatives in cosmetics, pharmaceutical products and foods. Parabens are esters of 4-hydroxybenzoic acid.

One common paraben used as a food preservative is ethylparaben.

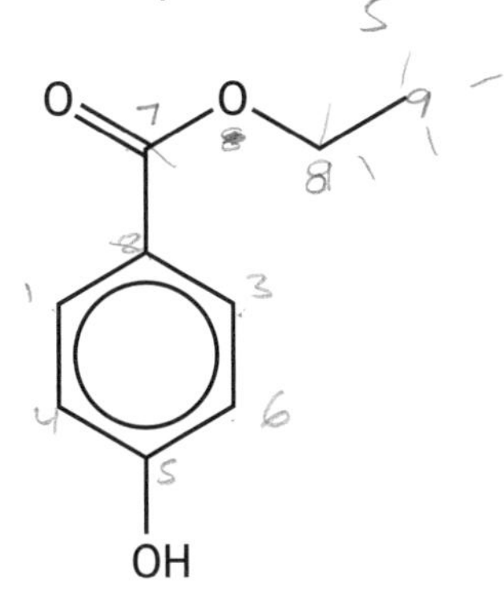


ethylparaben

(a) Ethylparaben is an aromatic compound containing both sigma and pi bonds.

(i) Write the molecular formula for ethylparaben. **1**

(ii) State the type of hybridisation which is adopted by the carbon atoms in the aromatic ring. **1**

(iii) Describe how pi bonds form. **1**

**[Turn over**

MARKS | DO NOT WRITE IN THIS MARGIN

**9. (continued)**

(b) Another preservative is sodium 4-hydroxybenzoate. It can be prepared by refluxing ethylparaben with sodium hydroxide solution.

(structure: O, O, OH) + $Na^+OH^-$ → (structure: O, $O^-Na^+$, OH) + $CH_3CH_2OH$

(i) Complete the diagram below to show how the reaction mixture is heated under reflux. **1**

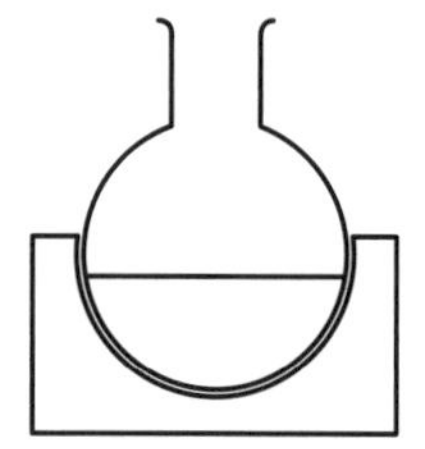

Heating mantle

(An additional diagram, if required, can be found on *Page 28*)

(ii) At the start of the reaction, two layers were observed in the reaction mixture.

Explain why only one layer was observed when the reaction was complete. **1**

MARKS | DO NOT WRITE IN THIS MARGIN

9. (b) (continued)

(iii) Explain fully why a solution of the salt sodium 4-hydroxybenzoate has a pH greater than 7. **2**

(iv) After refluxing, dilute hydrochloric acid was added to the reaction mixture and a white precipitate of 4-hydroxybenzoic acid was produced. The crude 4-hydroxybenzoic acid was recrystallised.

4-hydroxybenzoic acid is soluble in different solvents but only some of these solvents are suitable for recrystallisation.

State **two** factors that should be considered when selecting an appropriate solvent for this recrystallisation. **2**

(v) In this experiment, the percentage yield of 4-hydroxybenzoic acid was 77·5%.

Calculate the mass of ethylparaben (GFM = 166 g) required to produce 2·48 g of 4-hydroxybenzoic acid (GFM = 138 g). **2**

[Turn over

MARKS | DO NOT WRITE IN THIS MARGIN

**10.** Phenylbutazone is an anti-inflammatory drug used for the short-term treatment of pain and fever in animals.

(a) Phenylbutazone can be synthesised, in a multi-step process, starting from compound **A**.

Elemental microanalysis showed that compound **A** has a composition, by mass, of

50·0% C; 5·60% H; 44·4% O

Calculate the empirical formula of compound **A**. **2**

(b) An infra-red spectrum for compound **A** is shown below.

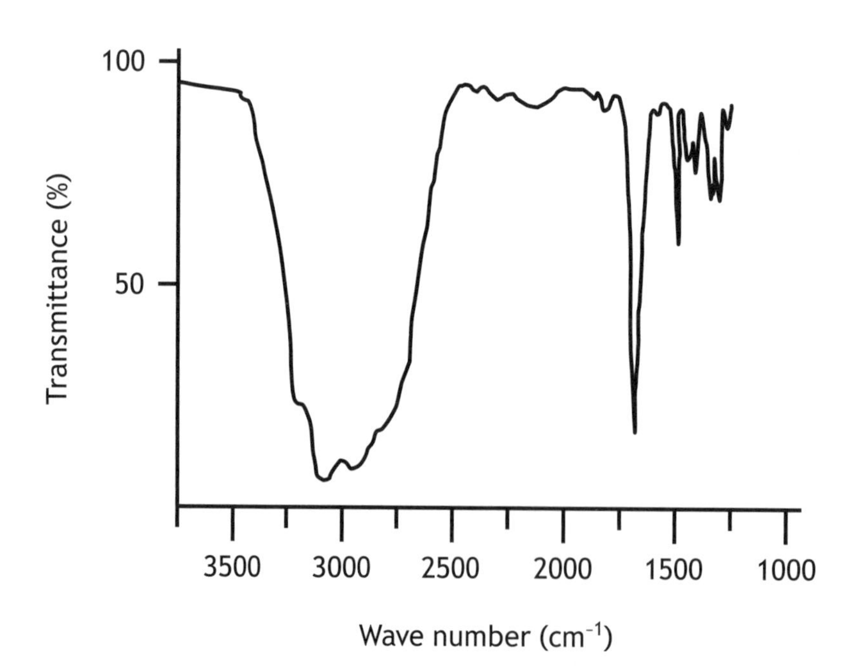


Identify the functional group responsible for the peak at 1710 $cm^{-1}$. **1**

MARKS | DO NOT WRITE IN THIS MARGIN

**10. (continued)**

(c) The mass spectrum for compound **A** is shown below.

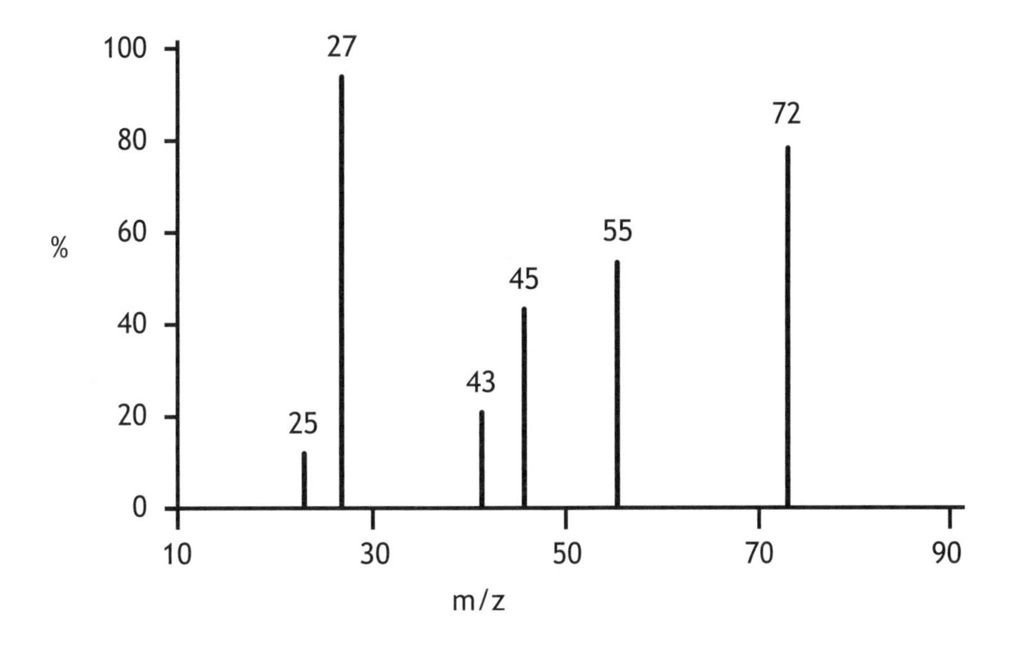


(i) Write the molecular formula for compound **A**. **1**

(ii) Suggest a possible ion fragment that may be responsible for the peak at m/z 27. **1**

(d) Considering all the evidence, draw a structural formula for compound **A**. **1**

**[END OF QUESTION PAPER]**

MARKS DO NOT WRITE IN THIS MARGIN

**ADDITIONAL DIAGRAM FOR USE IN QUESTION 3 (a) (i)**

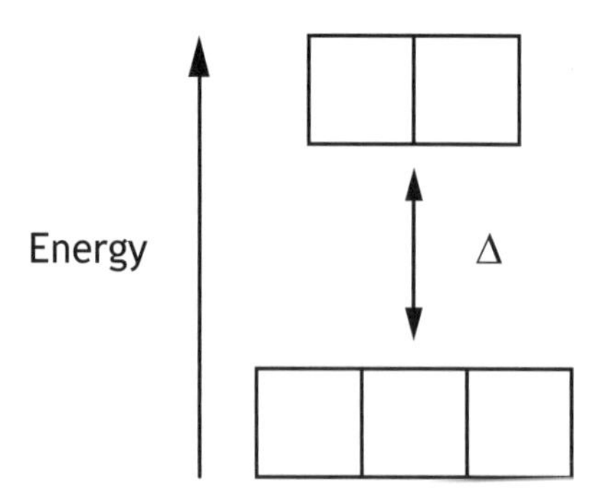


**ADDITIONAL DIAGRAM FOR USE IN QUESTION 9 (b) (i)**

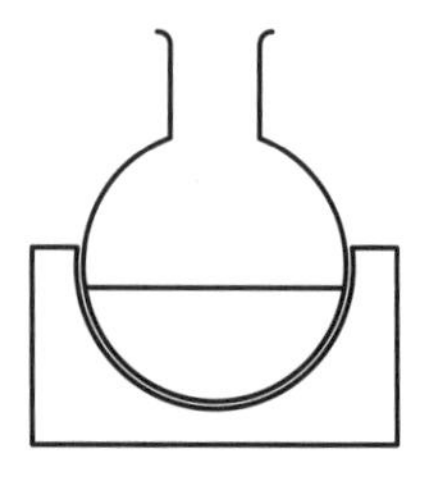

MARKS DO NOT WRITE IN THIS MARGIN

ADDITIONAL SPACE FOR ANSWERS AND ROUGH WORK

MARKS DO NOT WRITE IN THIS MARGIN

## ADDITIONAL SPACE FOR ANSWERS AND ROUGH WORK

ADVANCED HIGHER

# Answers

ANSWERS FOR

# SQA & HODDER GIBSON ADVANCED HIGHER CHEMISTRY 2016

## ADVANCED HIGHER CHEMISTRY
## 2015 SPECIMEN QUESTION PAPER

### SECTION 1

| Question | Response | Mark |
|---|---|---|
| 1. | D | 1 |
| 2. | A | 1 |
| 3. | C | 1 |
| 4. | B | 1 |
| 5. | C | 1 |
| 6. | B | 1 |
| 7. | D | 1 |
| 8. | D | 1 |
| 9. | D | 1 |
| 10. | B | 1 |
| 11. | C | 1 |
| 12. | D | 1 |
| 13. | A | 1 |
| 14. | B | 1 |
| 15. | C | 1 |
| 16. | B | 1 |
| 17. | D | 1 |
| 18. | D | 1 |
| 19. | B | 1 |
| 20. | A | 1 |
| 21. | D | 1 |
| 22. | A | 1 |
| 23. | C | 1 |
| 24. | C | 1 |
| 25. | A | 1 |
| 26. | B | 1 |
| 27. | C | 1 |
| 28. | A | 1 |
| 29. | D | 1 |
| 30. | A | 1 |

## SECTION 2

| Question | | | Expected response | Max mark | Additional guidance |
|---|---|---|---|---|---|
| 1. | (a) | | Lithium | 1 | |
| | (b) | | Energy absorbed resulted in electrons within atoms being promoted to higher energy level. **(1 mark)**<br>When the electron falls back to its original level, energy is emitted in the form of a photon whose wavelength corresponds to red light. **(1 mark)** | 2 | |
| | (c) | | $E = \frac{Lhc}{(1000)\lambda}$<br>or<br>$= \frac{6{\cdot}02 \times 10^{23} \times 6{\cdot}63 \times 10^{-34} \times 3 \times 10^{8}}{670{\cdot}7 \times 10^{-9} \times (1000)}$ **(1 mark)**<br>$= 179\,kJ\,mol^{-1}$ **(1 mark)** | 2 | 1000 may or may not be included at this step.<br>Significant figure rule applies.<br>Accept 178·53, 178·5 or 180. |
| 2. | (a) | | $ClO^-$ (aq),<br>or<br>hypochlorite ion | 1 | |
| | (b) | | $K_a = \frac{[H_3O^+][ClO^-]}{[HClO^-]}$ | 1 | allow $H^+$ |
| | (c) | | $ClO^-(aq) + H^+(aq) \rightleftharpoons HClO(aq)$<br>or<br>The $H^+(aq)$ are removed by the conjugate base from the water equilibrium **(1 mark)**<br>and<br>$H_2O(\ell) \rightleftharpoons H^+(aq) + OH^-(aq)$<br>or<br>This causes the water equilibrium to shift to the right-hand side producing excess $OH^-(aq)$ and hence pH $> 7$.<br>or<br>Produces an excess of hydroxide ions. **(1 mark)** | 2 | State symbols not required. |

| Question | | | Expected response | Max mark | Additional guidance |
|---|---|---|---|---|---|
| 3. | | | **The whole candidate response should first be read to establish its overall quality in terms of accuracy and relevance to the problem/situation presented.** There may be strengths and weaknesses in the candidate response: *assessors should focus as far as possible on the strengths, taking account of weaknesses (errors or omissions) only where they detract from the overall answer in a significant way, which should then be taken into account when determining whether the response demonstrates **reasonable, limited or no** understanding.*<br><br>**Assessors should use their professional judgement to apply the guidance below to the wide range of possible candidate responses.** | 3 | This open-ended question requires comment on **why β-carotene is orange.** Candidate responses are expected to make comment on the basis of relevant chemistry ideas/concepts which might include one or more of:<br><br>talk about conjugation, conjugated systems, chromophores, absorption of light, complementary colours, promotion of electrons, $\sigma$ and $\sigma^*$, $\pi$ and $\pi^*$, longer conjugation tends towards red, HOMO and LUMO. |
| | | | **3 marks:** The candidate has demonstrated a **good** conceptual understanding of the chemistry involved, providing a logically correct response to the problem/situation presented.<br><br>This type of response might include a statement of principle(s) involved, a relationship or equation, and the application of these to respond to the problem/situation. This does not mean the answer has to be what might be termed an "excellent" answer or a "complete" one. | | In response to this question, a **good** understanding might be demonstrated by a candidate response that:<br><br>• makes comments based on one relevant chemistry idea/concept, in a **detailed/developed** response that is **correct or largely correct** (any weaknesses are minor and do not detract from the overall response)<br><br>**or**<br><br>• makes comments based on a range of relevant chemistry ideas/concepts, in a response that is **correct or largely correct** (any weaknesses are minor and do not detract from the overall response)<br><br>**or**<br><br>• otherwise demonstrates a good understanding of the chemistry involved. |

| Question | | | Expected response | Max mark | Additional guidance |
|---|---|---|---|---|---|
| | | | **2 marks:** The candidate has demonstrated a **reasonable** understanding of the chemistry involved, showing that the problem/situation is understood.<br>This type of response might make some statement(s) that is/are relevant to the problem/situation, for example, a statement of relevant principle(s) or identification of a relevant relationship or equation. | | In response to this question, a **reasonable** understanding might be demonstrated by a candidate response that:<br>• makes comments based on one or more relevant chemistry idea(s)/concept(s), in a response that is **largely correct** but has **weaknesses** which detract to a small extent from the overall response<br>**or**<br>• otherwise demonstrates a reasonable understanding of the chemistry involved. |
| | | | **1 mark:** The candidate has demonstrated a **limited** understanding of the chemistry involved, showing that a little of the chemistry that is relevant to the problem/situation is understood.<br>The candidate has made some statement(s) that is/are relevant to the problem/situation. | | In response to this question, a **limited** understanding might be demonstrated by a candidate response that:<br>• makes comments based on one or more relevant chemistry idea(s)/concept(s), in a response that has **weaknesses** which detract to a large extent from the overall response<br>**or**<br>• otherwise demonstrates a limited understanding of the chemistry involved. |
| | | | **0 marks:** The candidate has demonstrated **no** understanding of the chemistry that is relevant to the problem/situation.<br>The candidate has made no statement(s) that is/are relevant to the problem/situation. | | Where the candidate has only demonstrated knowledge and understanding of chemistry **that is not relevant to the problem/situation presented**, 0 marks should be awarded. |
| 4. | (a) | (i) | +4, 4, 4+ , IV, four. | 1 | −4, 4− would not be accepted. |
| | | (ii) | [↑↓] [↑↓] [↑↓][↑↓][↑↓] [↑↓] [↑↓][↑↓][↑↓] [↑][↑][↑]<br>1s 2s 2p 3s 3p 3d | 1 | Box labels not required.<br>Noble gas start accepted in this case.<br>Full or half headed arrows accepted, but not vertical lines.<br>follow through |
| | | (iii) | The three d orbitals are all filled singly (with parallel spins). | 1 | follow through |
| | (b) | (i) | Degenerate means that the orbitals are of equal energy. | 1 | |

| Question | | | Expected response | Max mark | Additional guidance |
|---|---|---|---|---|---|
| | | (ii) | p-orbital<br>Z, Y, X | 1 | axes not necessary<br>any orientation accepted |
| | (c) | (i) | Several solutions of accurate permanganate concentration are made up and the absorbance of each is measured. **(1 mark)**<br>A calibration curve of concentration vs absorbance is drawn. **(1 mark)**<br>The absorbance of the solution of unknown permanganate concentration is measured and the calibration graph is then used to determine the concentration corresponding to this absorbance. **(1 mark)** | 3 | |
| | | (ii) | $[MnO_4^-] = 4{\cdot}25 \times 10^{-4}$ mol $l^{-1}$<br>moles $= 0{\cdot}25 \times 4{\cdot}25 \times 10^{-4}$<br>$= 1{\cdot}0625 \times 10^{-4}$ moles<br>moles Mn $= 1{\cdot}0625 \times 10^{-4}$ moles<br>mass of Mn $= 1{\cdot}0625 \times 10^{-4} \times 54{\cdot}9$ **(1 mark)**<br>$= 5{\cdot}833 \times 10^{-3}$g<br>% Mn $= 1{\cdot}7\%$ **(1 mark)** | 2 | 1 mark for concept of 54·9 × calculated moles of manganese.<br>1 mark for arithmetic.<br>Significant figure rule applies.<br>Accept 2, 1·67, 1·667. |
| 5. | (a) | | The mass must be in the region of 0·4 g and the exact mass must be known. | 1 | |
| | (b) | | Pour mixture into standard flask, rinse beaker with distilled water and add rinsings to standard flask. **(1 mark)**<br>and<br>Repeat, mix, make to mark with distilled water. **(1 mark)** | 2 | |
| | (c) | (i) | low GFM<br>or<br>unstable in air<br>or<br>absorbs moisture<br>or<br>is not a primary standard | 1 | |

| Question | | | Expected response | Max mark | Additional guidance |
|---|---|---|---|---|---|
| | | (ii) | Initial moles of HCl<br>0·02 × 1 = 0·02 mol<br><br>Ave titre = 12·65<br><br>No. moles NaOH =<br>0·1 × 0·01265 = 0·001265<br>= no. moles HCl left unreacted in 10 $cm^3$<br><br>10 × 0·001265 = 0·01265 in 100 $cm^3$ **(1 mark)**<br><br>Moles of acid reacting<br>0·02 − 0·01265 = 0·00735 mol **(1 mark)**<br><br>Mass of $CaCO_3$ is (0.00735/2) × 100·1<br>= 0·3679 g **(1 mark)**<br><br>% $CaCO_3$ (0·3679/0.390) × 100<br>= 94·3% **(1 mark)** | 4 | 1 mark concept of the scaling × 10.<br>1 mark concept of subtraction to calculate moles of acid reacting.<br><br>1 mark concept for stoichiometry 2 : 1.<br><br>1 mark for arithmetic.<br><br>Significant figure rule applies.<br><br>94, 94·33, 94·333. |
| 6. | (a) | | $C_{17}H_{19}N_3O_3S$ | 1 | Can be given in any order. |
| | (b) | | A racemate or racemic mixture. | 1 | |
| | (c) | | Antagonist | 1 | |
| | (d) | (i) | In an acidic environment, such as the stomach, the inactive isomer would be converted into the active one.<br><br>or<br><br>Cheaper to have a racemic mixture as you don't have to separate enantiomers. | 1 | |
| | | (ii) | The OH or hydroxyl group. | 1 | |
| 7. | (a) | | $\Delta S = \sum S°\text{products} - \sum S°\text{reactants}$<br>= (152 + 2 × 189) − (77 + 4 × 174) **(1 mark)**<br>= (530) − (773)<br>= −243 J $K^{-1}$ $mol^{-1}$ **(1 mark)** | 2 | |
| | (b) | | $\Delta G° = \Delta H° - T\Delta S°$<br>= (−244) − 298(−0·243) **(1 mark)**<br>= (−244) − (−72·414)<br>= −171·6 (kJ $mol^{-1}$) **(1 mark)**<br><br>Yes the reaction is feasible. **(1 mark)** | 3 | Working must be shown.<br><br>Follow through applies.<br><br>Units need not be given.<br><br>Answer in J acceptable. |
| 8. | (a) | (i) | Solvent extraction. | 1 | |
| | | (ii) | Drain layers into two separate beakers **(1 mark)**<br><br>Return lower/aqueous layer to separating funnel and add further diethylether and repeat **(1 mark)**<br><br>Evaporation/distillation of all diethylether layers **(1 mark)** | 3 | |
| | | (iii) | Immiscible in water **(1 mark)**<br><br>Benzocaine soluble in it **(1 mark)** | 2 | |

| Question | | | Expected response | Max mark | Additional guidance |
|---|---|---|---|---|---|
| | (b) | | recrystallisation<br>**or**<br>chromatography<br>**or**<br>glc | 1 | |
| | (c) | (i) | (Pure) benzocaine | 1 | |
| | | (ii) | Benzocaine has a small impurity.<br>**or**<br>Any other appropriate answer which suggests the benzocaine is not 100% pure. | 1 | |
| 9. | (a) | (i) | 2 | 1 | |
| | | (ii) | 1 | 1 | |
| | (b) | | Rate = $k[NO]^2[H_2]$ | 1 | |
| | (c) | | Using experiment 1,<br>$1{\cdot}20 \times 10^{-5} = k(4 \times 10^{-3})^2(1 \times 10^{-3})$<br>$k = \frac{1{\cdot}20 \times 10^{-5}}{(16 \times 10^{-6})(1 \times 10^{-3})}$<br>k = 750 **(1 mark)**<br>$mol^{-2}l^2s^{-1}$ **(1 mark)** | 2 | |
| 10. | (a) | | 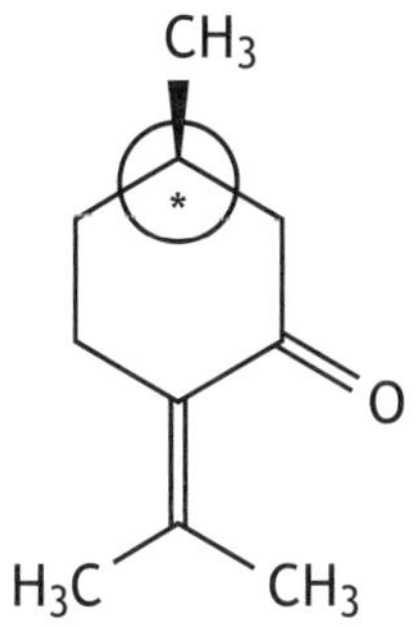 $CH_3$, $H_3C$, $CH_3$, O, * | 1 | |
| | (b) | | Infrared spectroscopy | 1 | |
| | (c) | | 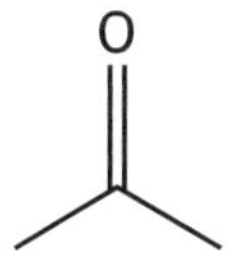 O **(1 mark)**<br>and 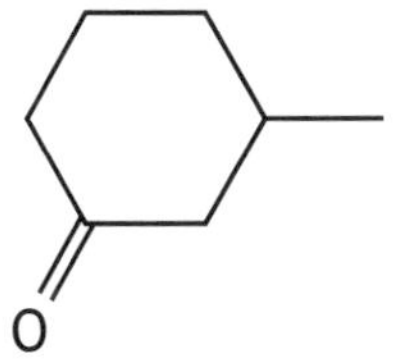<br>O **(1 mark)** | 2 | Correct full structural, shortened structural or skeletal formulae can all be accepted. |
| 11. | (a) | (i) | Cis refers to the geometric isomer where both substituents are on the same "side" of Pt. | 1 | |
| | | (ii) | A monodentate ligand forms one dative (covalent) bond (to a central metal atom or ion). | 1 | |

| Question | | | Expected response | Max mark | Additional guidance |
|---|---|---|---|---|---|
| | (b) | | Moles oxoplatin = $\frac{5 \cdot 00}{334 \cdot 1}$ = $1 \cdot 49656 \times 10^{-2}$<br>Moles asplatin = $1 \cdot 49656 \times 10^{-2}$<br>Mass asplatin = $1 \cdot 49656 \times 10^{-2} \times 484 \cdot 1$<br>= 7·245 g **(1 mark)**<br>% yield = $\frac{6 \cdot 36 \times 100}{7 \cdot 245}$ = 87·8% **(1 mark)** | 2 | Significant figure rule applies.<br>88, 87·78, 87·785. |
| 12. | | | **The whole candidate response should first be read to establish its overall quality in terms of accuracy and relevance to the problem/situation presented.** There may be strengths and weaknesses in the candidate response: *assessors should focus as far as possible on the strengths, taking account of weaknesses (errors or omissions) only where they detract from the overall answer in a significant way, which should then be taken into account when determining whether the response demonstrates* ***reasonable, limited*** *or* ***no*** *understanding.*<br><br>**Assessors should use their professional judgement to apply the guidance below to the wide range of possible candidate responses.** | 3 | This open-ended question requires candidates to outline the possible steps in the synthesis of propanoic acid from other small molecules such as ethene or ethanol. Candidate responses are expected to make comment on the basis of relevant chemistry ideas/concepts which might include one or more of: addition reaction, nucleophilic substitution reaction, reduction reaction, oxidation reaction and hydrolysis reaction. |
| | | | **3 marks:** The candidate has demonstrated a **good** conceptual understanding of the chemistry involved, providing a logically correct response to the problem/situation presented.<br><br>This type of response might include a statement of principle(s) involved, a relationship or equation, and the application of these to respond to the problem/situation. This does not mean the answer has to be what might be termed an “excellent” answer or a “complete” one. | | In response to this question, a **good** understanding might be demonstrated by a candidate response that:<br>• makes comments based on one relevant chemistry idea/concept, in a **detailed/developed** response that is **correct or largely correct** (any weaknesses are minor and do not detract from the overall response)<br>**or**<br>• makes comments based on a range of relevant chemistry ideas/concepts, in a response that is **correct or largely correct** (any weaknesses are minor and do not detract from the overall response)<br>**or**<br>• otherwise demonstrates a good understanding of the chemistry involved. |

| Question | | | Expected response | Max mark | Additional guidance |
|---|---|---|---|---|---|
| | | | **2 marks:** The candidate has demonstrated a **reasonable** understanding of the chemistry involved, showing that the problem/situation is understood.<br>This type of response might make some statement(s) that is/are relevant to the problem/situation, for example, a statement of relevant principle(s) or identification of a relevant relationship or equation. | | In response to this question, a **reasonable** understanding might be demonstrated by a candidate response that:<br>• makes comments based on one or more relevant chemistry idea(s)/concept(s), in a response that is **largely correct** but has **weaknesses** which detract to a small extent from the overall response<br>**or**<br>• otherwise demonstrates a reasonable understanding of the chemistry involved. |
| | | | **1 mark:** The candidate has demonstrated a **limited** understanding of the chemistry involved, showing that a little of the chemistry that is relevant to the problem/situation is understood.<br>The candidate has made some statement(s) that is/are relevant to the problem/situation. | | In response to this question, a **limited** understanding might be demonstrated by a candidate response that:<br>• makes comments based on one or more relevant chemistry idea(s)/concept(s), in a response that has **weaknesses** which detract to a large extent from the overall response<br>**or**<br>• otherwise demonstrates a limited understanding of the chemistry involved. |
| | | | **0 marks:** The candidate has demonstrated **no** understanding of the chemistry that is relevant to the problem/situation.<br>The candidate has made no statement(s) that is/are relevant to the problem/situation. | | Where the candidate has only demonstrated knowledge and understanding of chemistry **that is not relevant to the problem/situation presented**, 0 marks should be awarded. |
| 13. | (a) | (i) | Empirical formula<br>C H O<br>$\frac{79{\cdot}25}{12}$ $\frac{5{\cdot}66}{1}$ $\frac{15{\cdot}09}{16}$<br>$\frac{6{\cdot}60}{0{\cdot}94}$ $\frac{5{\cdot}66}{0{\cdot}94}$ $\frac{0{\cdot}94}{0{\cdot}94}$ **(1 mark)**<br>7 : 6 : 1<br>$C_7H_6O$ **(1 mark)** | 2 | 1 mark for calculating number of moles.<br>1 mark for formula. |
| | | (ii) | $C_7H_6O$ | 1 | |

| Question | | | Expected response | Max mark | Additional guidance |
|---|---|---|---|---|---|
| | (b) | | C=O<br>or<br>carbonyl group | 1 | |
| | (c) | | 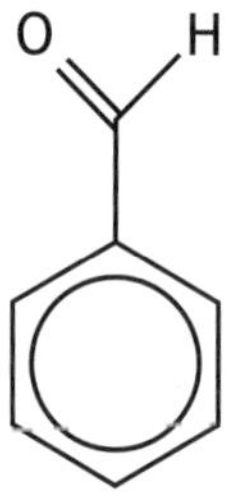 O H | 1 | Correct full structural, shortened structural or skeletal formulae can all be accepted.<br>Name is not required. |

# ADVANCED HIGHER CHEMISTRY MODEL PAPER 1

## SECTION 1

| Question | Response | Mark |
|---|---|---|
| 1. | D | 1 |
| 2. | D | 1 |
| 3. | C | 1 |
| 4. | C | 1 |
| 5. | A | 1 |
| 6. | A | 1 |
| 7. | C | 1 |
| 8. | B | 1 |
| 9. | C | 1 |
| 10. | A | 1 |
| 11. | B | 1 |
| 12. | A | 1 |
| 13. | B | 1 |
| 14. | A | 1 |
| 15. | B | 1 |
| 16. | B | 1 |
| 17. | A | 1 |
| 18. | D | 1 |
| 19. | C | 1 |
| 20. | B | 1 |
| 21. | A | 1 |
| 22. | B | 1 |
| 23. | C | 1 |
| 24. | D | 1 |
| 25. | B | 1 |
| 26. | C | 1 |
| 27. | B | 1 |
| 28. | D | 1 |
| 29. | D | 1 |
| 30. | D | 1 |

## SECTION 2

| Question | | | Expected response | Max mark | Additional guidance |
|---|---|---|---|---|---|
| 1. | (a) | | Energy is absorbed resulting in electrons being promoted to a higher energy level. **(1 mark)**<br><br>The line is produced from the electron falling to its original level, emitting energy which corresponds to the wavelength (or frequency) of one of the lines. **(1 mark)** | 2 | |
| | (b) | | $E - \frac{Lhc}{(1000)\,\lambda}$<br>or<br>$= \frac{6{\cdot}02 \times 10^{23} \times 6{\cdot}63 \times 10^{-34} \times 3 \times 10^{8}}{585 \times 10^{-9} \times (1000)}$ **(1 mark)**<br>$= 205\ \text{kJ mol}^{-1}$ **(1 mark)** | 2 | Alternatively, calculate the frequency first using $c = f\lambda$ and then calculate energy using $E = Lhf$.<br>i.e. $f = \frac{3 \times 108}{585 \times 10^{-9}}$<br>$f = 5{\cdot}13 \times 10^{14}\ s^{-1}$<br>$E = Lhf$<br>$= 6{\cdot}02 \times 10^{23} \times 6{\cdot}63 \times 10^{-34} \times 5{\cdot}13 \times 10^{14}$<br>$= 204\,680\ \text{J mol}^{-1}$ |
| | (c) | | 18 ppm = 18 mg per kg<br>10 kg = 180 mg of Ne = $180 \times 10^{-3}$ g **(1 mark)**<br>Moles = $\frac{180 \times 10^{-3}\ g}{20{\cdot}2} = 0{\cdot}00891$ mol **(1 mark)**<br>Volume = moles × molar volume<br>= 0·00891 × 22·4 litres<br>= 0·2 litres **(1 mark)** | 3 | Even if you forget the relationship between ppm and mass, you would still gain credit for attempting to calculate a mass (and hence, number of moles) and showing that you can convert moles into a volume. |
| 2. | (a) | | $H_2O(\ell) \rightleftharpoons H^+(aq) + OH^-(aq)$<br>Methanoate ions react with $H^+$ ions from water's equilibrium. **(1 mark)**<br>This causes the water equilibrium to shift to the right-hand side producing excess $OH^-$ and hence pH > 7. **(1 mark)**<br>**or**<br>Produces an excess of $OH^-$. | 2 | |
| | (b) | | $pH = pK_a - \log \frac{[acid]}{[salt]}$<br>$pH = pK_a - \log \frac{[0{\cdot}25]}{2{\cdot}5}$ **(1 mark)**<br>$= 3{\cdot}75 - (-1)$<br>$= 4{\cdot}75$ **(1 mark)** | 2 | |

| Question | | | Expected response | Max mark | Additional guidance |
|---|---|---|---|---|---|
| 3. | | | This is an open-ended question. Please refer to the specimen paper marking instructions. | 3 | **Suggestions from the author**<br>You could show your knowledge of chemistry by the following:<br>• discussing what is meant by ionisation energy and illustrate with equations<br>• explaining how increasing the nuclear charge pulls electrons closer to the nucleus<br>• using examples of elements in a period which follow the pattern suggested by the student<br>• discussing the exceptions to this rule and explaining in terms of their electronic configurations. For example:<br>Be $1s^22s^2$<br>B $1s^22s^22p^1$<br>B has a lower 1st IE than Be. Boron's outer electron is in a 2p orbital rather than a 2s. 2p is further from the nucleus than 2s. The increased distance results in a reduced attraction and so a reduced ionisation energy. In addition, the 2p orbital is screened by the $1s^2$ electrons AND by the $2s^2$ electrons. This reduces the pull from the nucleus and so lowers the ionisation energy. |
| 4. | (a) | | No. of moles of thiosulfate = 0·01525 × 0·102 = $1{\cdot}56 \times 10^{-3}$ so moles $Cu^{2+}$ = $1{\cdot}56 \times 10^{-3}$ **(1 mark)**<br><br>Mass Cu per sample = $63{\cdot}5 \times 1{\cdot}56 \times 10^{-3}$ = $9{\cdot}88 \times 10^{-2}$ **(1 mark)**<br><br>Mass of Cu in key = $9{\cdot}88 \times 10^{-2} \times \frac{1000}{25}$ = 3·95 g **(1 mark)** | 3 | Remember, at AH you must not round at intermediate steps, e.g. 3·87 g or 3·88 g. (2 out of 3 marks would be obtained from using 0·1 instead of 0·102.) |

| Question | | | Expected response | Max mark | Additional guidance |
|---|---|---|---|---|---|
| | (b) | | Any of the following:<br>• use distilled/deionised water<br>• rinsings<br>• carry out replicates/duplicates<br>• cover beaker with watch glass when key is being dissolved<br>• increase sample size for titration. | 1 | The following answers would receive zero:<br>• more titrations/more samples<br>• measure the 10 g of KI accurately<br>• use a more accurate balance. |
| | (c) | (i) | Hexaaquacopper(II) | 1 | Spelling must be correct. |
| | | (ii) | $[Cu(OH_2)_6]^{2+}$ octahedral structure: Cu bonded to six $OH_2$ ligands (two axial $OH_2$; equatorial $H_2O$, $OH_2$, $H_2O$, $OH_2$), in square brackets with 2+ charge | 1 | You must show that the copper bonds to the O of water. The square brackets, 2+ charge and octahedral shape must also be shown. |
| | | (iii) | An energy gap is created between the d orbitals **or** degeneracy is broken (when ligands bond). **(1 mark)**<br><br>This energy gap corresponds to a frequency of light which is absorbed. **(1 mark)**<br><br>Changing the ligand changes the energy gap between the d orbitals resulting in a different frequency of light being absorbed. **(1 mark)** | 3 | |
| 5. | (a) | | $\Delta S = \Sigma S^\circ$ products $- \Sigma S^\circ$ reactants<br>$= (213{\cdot}8 + 72{\cdot}1) - (112)$ **(1 mark)**<br>$= 173{\cdot}9\ J\ K^{-1}\ mol^{-1}$ **(1 mark)** | 2 | |
| | (b) | | $\Delta G^\circ = \Delta H^\circ - T\,\Delta S^\circ$<br>Just feasible when $\Delta G^\circ = 0$ **(1 mark)**<br>$\Delta H^\circ - T\,\Delta S^\circ = 0$<br>$T\,\Delta S^\circ = \Delta H^\circ$<br>$T = \frac{266\,000}{173{\cdot}9}$ **(1 mark)**<br>$= 1529{\cdot}6\ K$ **(1 mark)** | 3 | Since $\Delta H^\circ$ is in kJ and $\Delta S^\circ$ is in J, you must change either the enthalpy or entropy so that you are using the same energy units. In the marking scheme, the enthalpy has been ×1000 so that it has units in J. |
| 6. | (a) | | +5 or 5 | 1 | 3 × −2 for O = −6.<br>Since the overall charge is −1, Br must be +5. |
| | (b) | (i) | 1st order for $Br^-$ **(1 mark)**<br>2nd order for $H^+$ **(1 mark)** | 2 | |
| | | (ii) | Rate = $k[BrO_3^-][Br^-][H^+]^2$ | 1 | k must be lowercase since capital K represents the units of temperature (Kelvin). |

| Question | | | Expected response | Max mark | Additional guidance |
|---|---|---|---|---|---|
| | | (iii) | 8 **(1 mark)**<br>Units: $l^3\,mol^{-3}\,s^{-1}$ **(1 mark)** | 2 | The answer for k is obtained by rearranging the equation in (ii) and substituting the data from any of the experiments. |
| 7. | (a) | | Hexa-2,4-diene | 1 | |
| | (b) | | Any correct structure for deca-2,4,6,8-tetraene such as 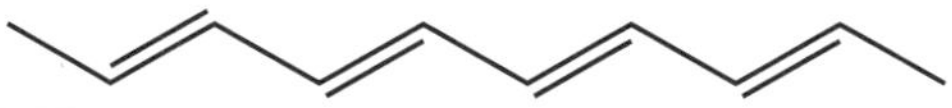 | 1 | |
| | (c) | | Conjugated/conjugation | 1 | |
| | (d) | | As the number of alternating single and double bonds increases, the energy difference between the HOMO and the LUMO decreases. | 1 | |
| | (e) | | α-carotene has a shorter conjugated system or similar. | 1 | |
| | (f) | | 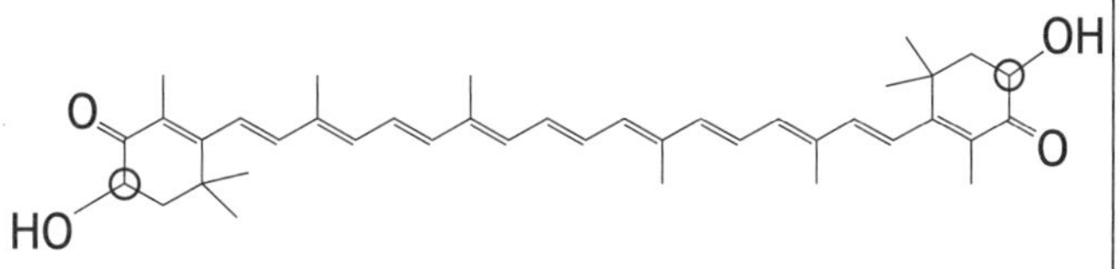 O OH O HO<br>Either of the two carbons connected to an OH group<br>**or**<br>section with one of the asymmetric carbon atoms circled. | 1 | Look for a C atom that is connected to four different atoms/groups of atoms. |
| 8. | (a) | | Ether | 1 | |
| | (b) | | Add an alkali metal to methanol | 1 | |
| | (c) | (i) | First step – show the heterolytic fission of the C—Cl bond to form the carbocation. 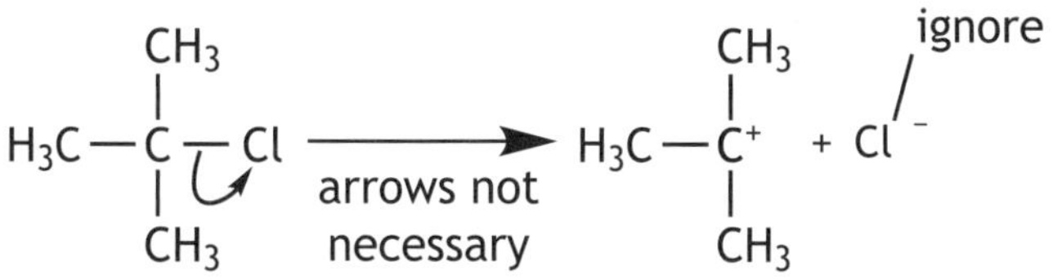<br>$H_3C—C(CH_3)_2—Cl \rightarrow H_3C—C^+(CH_3)_2 + Cl^-$ (arrows not necessary; ignore)<br>Second step – show the nucleophilic attack of the methoxide ion. 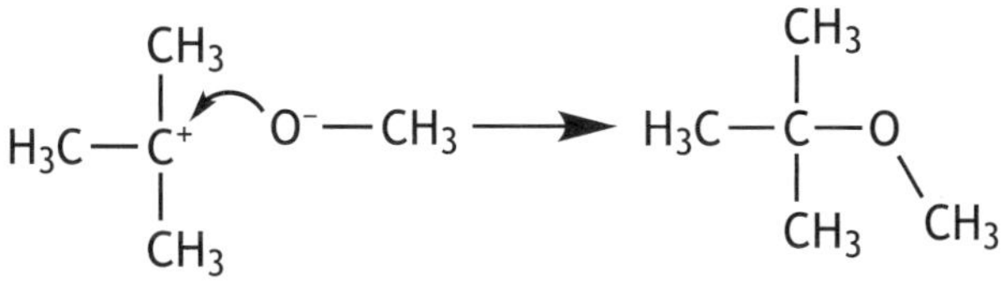<br>$H_3C—C^+(CH_3)_2 \;\; O^-—CH_3 \rightarrow H_3C—C(CH_3)_2—O—CH_3$<br>(**1 mark** for correct reactant and for curly arrow showing heterolytic fission of C—Cl bond<br>**1 mark** for correct carbocation intermediate<br>**1 mark** for second curly arrow and final product) | 3 | |

| Question | | | Expected response | Max mark | Additional guidance |
|---|---|---|---|---|---|
| | | (ii) | Should mention one of two concepts:<br>**either**<br>the carbocation formed has three methyl (alkyl) groups attached which can feed in electron density stabilising the positive charge thus making an $S_N1$ mechanism more favourable<br>**or**<br>the tertiary haloalkane has three methyl groups attached which offer steric hindrance w.r.t. the formation of the five co-ordinate transition state seen as part of the $S_N2$ mechanism; bulky groups/too crowded/steric hindrance. | 1 | |
| | (d) | (i) | An equal mixture of enantiomers/both optical isomers | 1 | |
| | | (ii) | Nothing | 1 | One enantiomer would rotate the polarised light in one direction; the other enantiomer would rotate the light in the opposite direction. As there is an equal amount of each enantiomer, the rotations would cancel. |
| 9. | (a) | | Fairly high molecular mass/available in high purity/thermodynamically stable/soluble in water | 1 | |
| | (b) | | 1. Calculate the mass needed.<br>2. Accurately weigh required mass of $KIO_3$ and dissolve completely in small volume of water.<br>3. Transfer the solution to a standard flask, rinsing the beaker with deionised water and transferring the rinsings to the flask.<br>4. Add deionised water up to the mark adding the last few drops with a dropper.<br>5. Invert to mix. | 2 | |
| | (c) | | Moles of $KIO_3$ required = CV = 0·1 × 0·1 = 0·01<br>Volume of 1·00 mol $l^{-1}$ solution required<br>$= \frac{\text{moles}}{C}$<br>$= \frac{0{\cdot}01}{1} = 0{\cdot}01$ litres<br>Measure 10 $cm^3$ of 1·00 mol $l^{-1}$ stock solution using a pipette and transfer to a 100 $cm^3$ standard flask. Add deionised water up to the mark, stopper and invert.<br>**(1 mark for correct volume; 1 mark for correct procedure)** | 2 | |

| Question | | | Expected response | Max mark | Additional guidance |
|---|---|---|---|---|---|
| 10. | (a) | | Condensation | 1 | |
| | (b) | | $CH_3CH_2CN$<br>or<br>H–C(H)(H)–C(H)(H)–C≡N | 1 | |
| | (c) | | H–C(H)(H)–C(H)(H)–C(=O)–O–C(H)(H)–H<br>Hc is the three H atoms on the left-hand side of the molecule. | 1 | |

| Question | | | Expected response | Max mark | Additional guidance |
|---|---|---|---|---|---|
| | (d) | | This is an open-ended question. Please refer to the specimen paper marking instructions. | 3 | **Suggestions from the author**<br>IR: Discuss how IR can identify functional groups. Quote some of the wavenumbers from the data booklet for the compounds of interest, i.e. for conversion from an alcohol and carboxylic acid into an ester you would expect to see a decrease/disappearance of —OH from the alcohol and acid and an appearance of the ester functional group.<br>This could be done by sampling at fixed periods of time and observing the change in IR spectrum as the reaction proceeds.<br>Could discuss comparing the reaction mixture with reference samples/library spectra of the product ester.<br>Chromatography: Could run a TLC of three standards, i.e. methanol, propanoic acid and methyl propanoate to calculate/observe their Rf. Run plates with these three spots alongside samples from the reaction to see the appearance of a spot corresponding to the ester (compare with the reference).<br>Discuss the need for spotting agents.<br>Discuss the pros and cons of each technique. |

| Question | | | Expected response | Max mark | Additional guidance |
|---|---|---|---|---|---|
| 11. | (a) | | Condensor<br>Heating mantle | 2 | |
| | (b) | (i) | To purify (the sulfanilamide)<br>To get rid of impurities | 1 | |
| | | (ii) | Insoluble in cold water<br>Soluble in hot water | 1 | |
| | | (iii) | Vaccuum filtration | 1 | |
| | (c) | | GFM reactant 214·1, GFM product 172·1<br>Both GFM calculated correctly **(1 mark)**<br>$\frac{4{\cdot}282}{214{\cdot}1} \times 172{\cdot}1 = 3{\cdot}442$ **(1 mark)**<br>$\frac{2{\cdot}237}{3{\cdot}442} \times 100 = \mathbf{65\%}$ **(1 mark)**<br>**or** $\frac{4{\cdot}282}{214.1} = 0{\cdot}02$ moles $\times 172{\cdot}1 = 3{\cdot}442$ **(2 marks)**<br>Then $\frac{2{\cdot}237}{3{\cdot}442} \times 100 = 65\%$ **(1 mark)** | 3 | |
| | (d) | | The sample is mixed with pure sulfanilamide **(1 mark)**<br>The melting point of the mixture will be the same **(1 mark)** | 2 | |
| | (e) | | Chromatography or IR or NMR or TLC | 1 | |

# ADVANCED HIGHER CHEMISTRY MODEL PAPER 2

## SECTION 1

| Question | Response | Mark |
|---|---|---|
| 1. | A | 1 |
| 2. | A | 1 |
| 3. | D | 1 |
| 4. | D | 1 |
| 5. | B | 1 |
| 6. | B | 1 |
| 7. | B | 1 |
| 8. | C | 1 |
| 9. | C | 1 |
| 10. | D | 1 |
| 11. | D | 1 |
| 12. | C | 1 |
| 13. | D | 1 |
| 14. | C | 1 |
| 15. | D | 1 |
| 16. | D | 1 |
| 17. | B | 1 |
| 18. | B | 1 |
| 19. | A | 1 |
| 20. | A | 1 |
| 21. | C | 1 |
| 22. | A | 1 |
| 23. | C | 1 |
| 24. | A | 1 |
| 25. | C | 1 |
| 26. | D | 1 |
| 27. | D | 1 |
| 28. | C | 1 |
| 29. | C | 1 |
| 30. | B | 1 |

## SECTION 2

| Question | | | Expected response | Max mark | Additional guidance |
|---|---|---|---|---|---|
| 1. | (a) | | $PH_3$ structure: P bonded to three H atoms (one wedge, one hashed bond) | 1 | |
| | (b) | (i) | $n = 3$, $l = 0$, $m = 0$, $s = +\frac{1}{2}$ or $-\frac{1}{2}$ | 2 | The first quantum number ($n$) refers to the shell the electron is in. In this case, it is in the 3rd shell so $n = 3$. The second quantum number ($l$) refers to the orbital, i.e. is it an s, p, d etc. In this case, it is an s orbital which is known as $l = 0$. If it was a p orbital, $l$ would be 1; if it was a d orbital, $l$ would be 2. The third quantum number ($m$) for an s orbital is always 0. For p orbitals, $m$ can be −1, 0 and +1 (px, py and pz). The final quantum number ($s$) refers to the spin of the electron. It is either $+\frac{1}{2}$ or $-\frac{1}{2}$. |
| | | (ii) | The 1s orbital has the same shape as the 2s orbital but is smaller<br>**or**<br>has a lower energy than the 2s orbital. | 1 | |
| | | (iii) | The two electrons occupy degenerate 3p orbitals singly and with parallel spins. | 1 | |
| | | (iv) | 4d | 1 | |
| | (c) | | 5000 ppm = 5000 mg per kg<br>2 kg = 10 000 mg of $CH_4$ = 10 g **(1 mark)**<br>Moles = $\frac{10\,g}{16}$ = 0·625 mol **(1 mark)** | 2 | |
| 2. | (a) | (i) | Moles EDTA = 0·11 × 0·001175 = $1{\cdot}29 \times 10^{-3}$<br>Moles Ni = 4 × moles EDTA = $5{\cdot}16 \times 10^{-3}$ **(1 mark)**<br>Mass Ni = $5{\cdot}16 \times 10^{-3} \times 58{\cdot}7$ = 0·303 g **(1 mark)**<br>% Ni = $\frac{0{\cdot}303}{1{\cdot}33} \times 100$ = 22·8% **(1 mark)** | 3 | 1 mole of EDTA reacts with 1 mole of the Ni ions. The number of moles calculated has to be ×4 since only 25 $cm^3$ of the original 100 $cm^3$ solution was used. |

| Question | | | Expected response | Max mark | Additional guidance |
|---|---|---|---|---|---|
| | | (ii) | 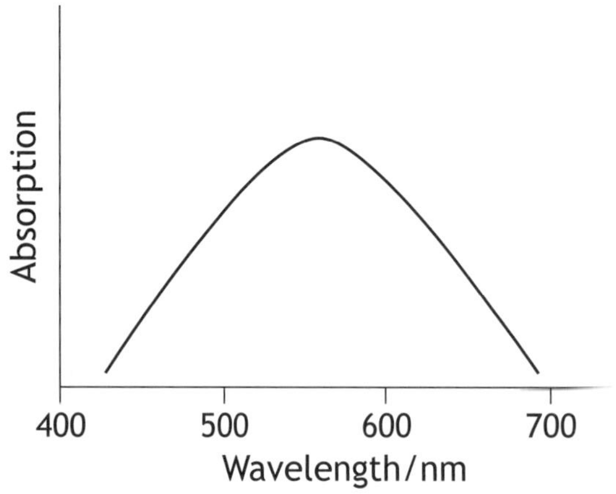 Absorption<br>400 500 600 700<br>Wavelength/nm | 1 | To produce a violet colour, red and blue must be mixed, i.e the sample must not absorb red or blue. Assuming the three colours which mix to make white light are red, green and blue, the sample must absorb strongly in the green (509 nm) region but allow all red (~650 nm) and blue (~480 nm) to transmit.<br><br>The data booklet can be used to determine the approximate $\lambda$ of red, green and blue. |
| | (b) | | Titration: end point uncertainty **(1 mark)**<br>Colorimetry: reliability of calibration graph or dilute solutions giving low absorbance **(1 mark)**<br>Gravimetric: small mass of sample giving large uncertainty in balance readings/mass transfer or mechanical losses **(1 mark)** | 3 | These are the sources of most significant error. You may be tempted to mention apparatus error, but this is not a major factor except for gravimetric analysis where the ability of the balance to measure accurate masses is crucial. |
| | (c) | | $E = \frac{Lhc}{(1000)\,\lambda}$<br>**or**<br>$= \frac{6 \cdot 02 \times 10^{23} \times 6 \cdot 63 \times 10^{-34} \times 3 \times 10^{8}}{508 \times 10^{-9} \times (1000)}$ **(1 mark)**<br>$= 236\ kJ\ mol^{-1}$ **(1 mark)** | 2 | Alternatively, calculate the frequency first using $c = f\lambda$ and then calculate energy using $E = Lhf$.<br><br>i.e. $f = \frac{3 \times 10^{8}}{508 \times 10^{-9}}$<br>$f = 5 \cdot 91 \times 10^{14}\ s^{-1}$<br><br>$E = Lhf$<br>$= 6 \cdot 02 \times 10^{23}$<br>$\times 6 \cdot 63 \times 10^{-34}$<br>$\times 5 \cdot 91 \times 10^{14}$<br>$= 235\,704\ J\ mol^{-1}$ |

| Question | | | Expected response | Max mark | Additional guidance |
|---|---|---|---|---|---|
| 3. | | | This is an open-ended question.<br>Refer to the marking scheme from the specimen paper. | 3 | **Suggestions from the author**<br>You should use lots of examples to illustrate your understanding of colour. You could discuss:<br>• the colour in transition metal compounds caused by loss of d orbital degeneracy<br>• the conjugation in organic molecules causing electron movement between HOMO and LUMO<br>• the emission of photons from atoms as a result of energy being absorbed which corresponds to the energy gap in an atom.<br>If you can quote specific examples of coloured compounds (and the reasons for the colour), use them! |
| 4. | (a) | | $K = \frac{[CH_4][H_2S]^2}{[CS_2][H_2]^4}$ | 1 | |
| | (b) | | 281 | 1 | |
| 5. | (a) | | $HCO_3^-(aq)$<br>**or**<br>$HCO_3^-$ | 1 | |
| | (b) | | $pH = \frac{1}{2}pK_a - \frac{1}{2}\log c$<br>Substitute values: $pH = \frac{1}{2}(6{\cdot}35) - \frac{1}{2}\log 0{\cdot}01$ **(1 mark)**<br>pH = 4·18 **(1 mark)** | 2 | |
| | (c) | (i) | Any indicator with a pH change in the alkaline region. From the data booklet, the following would be acceptable: phenolphthalein, cresol red, thymol blue or thymolphthalein. | 1 | The solution formed at the end point would contain potassium carbonate which is an alkaline salt. |
| | | (ii) | **1 mark** for stating that a buffer is formed.<br>**1 mark** for explaining that excess $H^+$ ions are removed by the conjugate base (formed from the potassium carbonate) **or** that excess base reacts with $H^+$ from the weak acid. | 2 | A weak acid mixed with the salt of a weak acid forms a buffer solution. |
| 6. | (a) | | $-220{\cdot}3\ J\ K^{-1}\ mol^{-1}$ | 1 | $\Delta S = \sum S_{products} - \sum S_{reactants}$<br>$= ((2 \times 27) + 5{\cdot}7) - ((2 \times 33) + 214)$<br>$= (59{\cdot}7) - (280)$<br>$= -220{\cdot}3$ |

| Question | | | Expected response | Max mark | Additional guidance |
|---|---|---|---|---|---|
| | (b) | | $\Delta H^o = -592\ kJ\ mol^{-1}$ **(1 mark)**<br>$\Delta G^o = \Delta H^o - T\Delta S^o$<br>$= -592 - 298 \times \left(\frac{-220 \cdot 3}{1000}\right)$ **(1 mark)**<br>$= -526 \cdot 35\ kJ\ mol^{-1}$ ($-526\ kJ\ mol^{-1}$) **(1 mark)** | 3 | To obtain ΔH, use Hess' law to rearrange the equations given,<br>i.e. $\Delta H = (2 \times -493) + 394$.<br>Since this is standard conditions, T = 298 K.<br>Remember to divide the ΔS by 1000 so that all units are in kJ. |
| 7. | (a) | | 3rd order | 1 | This is obtained by adding the indices from the rate equation, i.e. 2 (from NO) and 1 (from $Cl_2$). |
| | (b) | | Reaction 3, since the rate equation shows that the rate is independent of the concentration of ammonia. | 1 | In other words, it doesn't matter what the concentration of ammonia is, the rate is not affected. |
| | (c) | | $k = \frac{[Rate]}{[NO]^2[Cl_2]} = \frac{1 \cdot 43 \times 10^{-6}}{[0 \cdot 250]^2[0 \cdot 250]}$<br>$= 9 \cdot 15 \times 10^{-5}\ l^2\ mol^{-2}\ s^{-1}$<br>**(1 mark for correct value; 1 mark for correct units)** | 2 | |
| 8. | (a) | | Moles sulfuric = $0 \cdot 1 \times 0 \cdot 0121 = 1 \cdot 21 \times 10^{-3}$<br>Moles of NaOH in 25 $cm^3$ = $1 \cdot 21 \times 10^{-3} \times 2$<br>$= 2 \cdot 42 \times 10^{-3}$ **(1 mark)**<br>Moles of NaOH in 500 $cm^3$ = $20 \times 2 \cdot 42 \times 10^{-3}$<br>$= 0 \cdot 0484$ **(1 mark)**<br>Moles NaOH added at the start = $1 \times 0 \cdot 050$<br>$= 0 \cdot 05$<br>Moles NaOH reacting = $0 \cdot 05 - 0 \cdot 0484$<br>$= 1 \cdot 6 \times 10^{-3}$ **(1 mark)**<br>Moles of aspirin = $\frac{1 \cdot 6 \times 10^{-3}}{2} = 8 \times 10^{-4}$<br>Mass of aspirin = $8 \times 10^{-4} \times 180$<br>$= 0 \cdot 144$ g **(1 mark)** | 4 | |
| | (b) | | Aspirin is not very soluble in water. | 1 | |
| | (c) | (i) | Any of the following for **1 mark:**<br>• mass transfer or mechanical losses<br>• purification of the product results in loses<br>• side reactions<br>• the aspirin hydrolyses to form reactants<br>• impure reactants | 1 | |

| Question | | | Expected response | Max mark | Additional guidance |
|---|---|---|---|---|---|
| | | (ii) | 74% of x = 8 g<br>$x = \frac{8}{0{\cdot}74} = 10{\cdot}810\,g$ **(1 mark)**<br>Moles of aspirin = $\frac{10{\cdot}810}{180} = 0{\cdot}060$<br>Moles of 2-hydroxybenzoic acid = 0·060<br>Mass of 2-hydroxybenzoic acid = 0·060 × 138<br>= 8·29 g<br>**(1 mark)** | 2 | |
| | | (iii) | **1 mark** for melting point.<br>**1 mark** for TLC.<br><br>**Melting point:**<br>• Compare the melting point obtained with the literature values. A sharp mp close to the literature value would suggest high purity.<br>• A different mp would suggest a different compound.<br>• A lower mp over a long range would suggest the presence of impurities.<br><br>**TLC:**<br>• A single spot would suggest 1 compound; more than 1 spot would suggest impurities.<br>• Run the TLC with a control (pure sample of aspirin). $R_f$ values for the control should match the sample made. | 2 | |
| | (d) | (i) | It means that you need to know the exact (accurate) mass of the compound you weigh and it should be in the region of 0.40 g. | 1 | |
| | | (ii) | 100 $cm^3$ flask contains 0·4 g<br>500 $cm^3$ flask contains 0·4 g<br><br>Concentration = $\frac{mass}{volume} = \frac{0{\cdot}4}{0{\cdot}5} = 0{\cdot}8\,gl^{-1}$ | 1 | |
| | | (iii) | Choose a filter where maximum absorbance occurs. | 1 | |
| | | (iv) | Use a pipette to measure out the 10 $cm^3$ instead of a burette. | 1 | |

| Question | | | Expected response | Max mark | Additional guidance |
|---|---|---|---|---|---|
| 9. | (a) | | Phosphoric acid **or** sulfuric acid **or** aluminium oxide | 1 | |
| | (b) | | This is an open-ended question. Please refer to the specimen paper marking instructions. | 3 | **Suggestions from the author**<br>Cyclohexanone is a ketone. You could carry out chromatography, comparing the retention time of the product with that of a pure sample of cyclohexanone.<br>Spectroscopic techniques would give lots of information:<br>Mass spec would give an ion fragment pattern which you could compare to the formula mass of this product.<br>$^{1}$H NMR would give a distinctive splitting pattern (you could discuss a likely pattern) or you could compare a spectrum of a pure (or reference) sample of cyclohexanone with this sample, likewise with all spectroscopic techniques.<br>IR-specific carbonyl stretch – quote the data booklet value.<br>Empirical formula – determination from microanalysis. |
| 10. | (a) | | Electrophile | 1 | |
| | (b) | (i) | H; C; $H_3C$; $CO_2^-$; OH or H; C; HO; $CH_3$; $CO_2^-$ or equivalent<br>Must be tetrahedral but dots and wedges can be replaced by solid lines. | 1 | |

| Question | | | Expected response | Max mark | Additional guidance |
|---|---|---|---|---|---|
| | | (ii) | While one group would be able to bind to the appropriate region, the other two would not.<br>**or**<br>The three "functional" groups fail to match the binding regions of the active site.<br>**or**<br>Only one group or two groups could bind (or match) the binding regions.<br>**or**<br>The groups on the lactate ion no longer match the binding regions on the active site of the enzyme.<br>**or**<br>The lactate ion no longer complements the binding region (of the active site).<br>**or**<br>The groups now fail to match the binding region (of the active site). | 1 | |
| 11. | (a) | | HBr **or** hydrogen bromide | 1 | |
| | (b) | (i) | Mixing of the s orbital with two p orbitals | 1 | |
| | | (ii) | Sigma bonds – end-on overlap of (atomic) orbitals<br>Pi bonds – sideways overlap of (atomic) orbitals | 1 | |
| | (c) | | 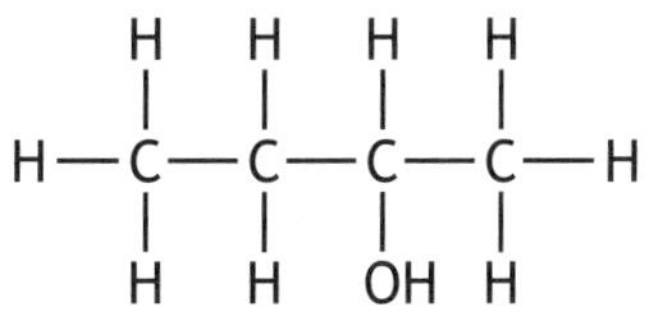 H—C(H)(H)—C(H)(H)—C(H)(OH)—C(H)(H)—H or $CH_3CH_2CHOHCH_3$ | 1 | |
| | (d) | | 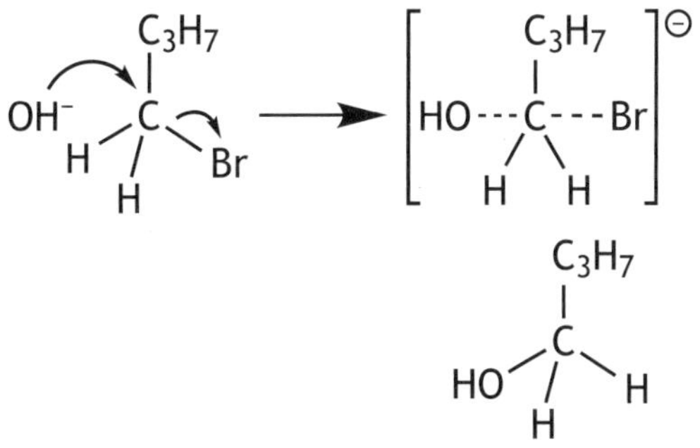 $OH^-$ attacks C (bonded to $C_3H_7$, H, H, Br); Br leaves → $[HO\text{---}C(C_3H_7)(H)(H)\text{---}Br]^{\ominus}$ → $HO\text{–}C(C_3H_7)(H)(H)$<br>–ve charge must be given outside the brackets or on the C (in the transition state) | 2 | |

| Question | | | Expected response | Max mark | Additional guidance |
|---|---|---|---|---|---|
| 12. | (a) | (i) | 3·52 g $CO_2$ contains 0·96 g C<br>1·44 g $H_2O$ contains 0·16 g H<br>**(both for 1 mark)**<br>Therefore, 0·64 g O **(1 mark)** | 2 | |
| | | (ii) | $C = \frac{0 \cdot 96}{12} = 0 \cdot 08$; $H = \frac{0 \cdot 16}{1} = 0 \cdot 16$;<br>$O = \frac{0 \cdot 64}{16} = 0 \cdot 04$<br>Ratio = 2 : 4 : 1<br>Empirical formula is therefore $C_2H_4O$. | 1 | |
| | (b) | | $C_4H_8O_2$ | 1 | |
| | (c) | (i) | $CH_3-CH_2-CH_2-COOH$ (full structural formula: H—C(H)(H)—C(H)(H)—C(H)(H)—C(=O)—O—H)<br>Butanoic acid | 2 | |
| | | (ii) | Acts as a reference | 1 | |

# ADVANCED HIGHER CHEMISTRY 2016

## SECTION 1

| Question | Answer | Max Mark |
|---|---|---|
| 1. | C | 1 |
| 2. | B | 1 |
| 3. | C | 1 |
| 4. | D | 1 |
| 5. | C | 1 |
| 6. | A | 1 |
| 7. | C | 1 |
| 8. | A | 1 |
| 9. | A | 1 |
| 10. | C | 1 |
| 11. | B | 1 |
| 12. | D | 1 |
| 13. | C | 1 |
| 14. | D | 1 |
| 15. | B | 1 |
| 16. | D | 1 |
| 17. | A | 1 |
| 18. | D | 1 |
| 19. | A | 1 |
| 20. | C | 1 |
| 21. | C | 1 |
| 22. | D | 1 |
| 23. | A | 1 |
| 24. | C | 1 |
| 25. | D | 1 |
| 26. | B | 1 |
| 27. | D | 1 |
| 28. | B | 1 |
| 29. | A | 1 |
| 30. | B | 1 |

## SECTION 2

| Question | | | Expected response | Max mark | Additional guidance |
|---|---|---|---|---|---|
| 1. | (a) | (i) | −44 kJ mol$^{-1}$ | 1 | −40 also acceptable<br><br>Units not needed but must be correct if given. |
| | | (ii) | −130 J K$^{-1}$mol$^{-1}$ (3)<br><br>**or**<br><br>ΔG° = ΣG°products − ΣG°reactants<br>= −175 − (68 − 237)<br>= −6 (kJ mol$^{-1}$) (1)<br><br>ΔS° = (ΔH° − ΔG°)/T<br>(**or** use of ΔG° = ΔH° − TΔS°) (1)<br><br>= (−44 − (−6))/298<br>= −0·128 kJ K$^{-1}$mol$^{-1}$<br>= −130 J K$^{-1}$mol$^{-1}$ (1) | 3 | −100/−128/−127·5 also acceptable.<br><br>Units not needed for final answer but must be correct if given.<br><br>Follow through applies. |
| | (b) | | 340 K (2)<br><br>**or**<br><br>Reaction becomes feasible when<br>ΔG = 0 (1)<br><br>Therefore T = ΔH/ΔS<br><br>T = 44/0·130<br>T = 340 K (1) | 2 | 300/338/338·5 are also acceptable.<br><br>Follow through applies, from **(a)(i)** and/**or (a)(ii)**.<br><br>Units not needed for final answer but must be correct if given. |
| 2. | (a) | | 1s$^2$ 2s$^2$ 2p$^6$ | 1 | If orbital boxes are given they must be correct and the correct notation is also required. |
| | (b) | |  | 1 | Any orientation of this shape allowed.<br><br>If axes are drawn, then a lobe of the orbital must lie on an axis. |
| | (c) | | −1, 0, (+)1 | 1 | |

| Question | | | Expected response | Max mark | Additional guidance |
|---|---|---|---|---|---|
| 3. | (a) | (i) | 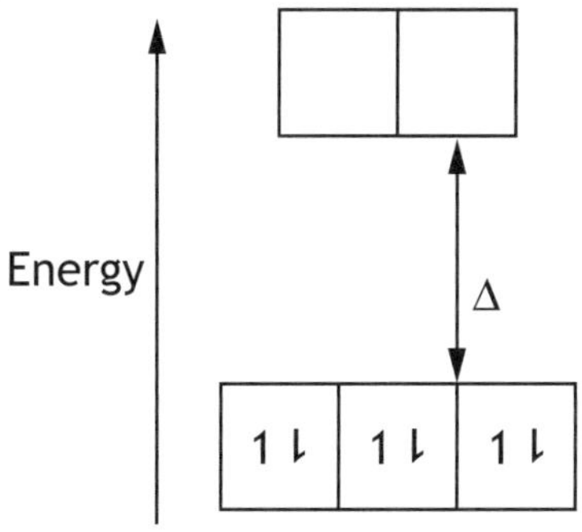 | 1 | Full headed arrows are acceptable, but spin must be shown. |
| | | (ii) | 1 mark is awarded for recognising that there is a small energy gap.<br><br>1 mark is awarded for recognising that electrons are promoted. | 2 | Less energy is required to promote an electron. **(2)**<br><br>**or**<br><br>Δ is small/less (than for $CN^-$).<br><br>**or**<br><br>Energy difference between levels is less. **(1)**<br><br>Electrons can occupy all of the d-orbitals.<br><br>**or**<br><br>Electrons can occupy the higher energy d-orbitals.<br><br>**or**<br><br>Electrons can be promoted between energy levels. **(1)** |
| | | (iii) | $Fe^{3+}$ has five/odd number of (d-)electrons.<br><br>**or**<br><br>It is $3d^5$.<br><br>**or**<br><br>It has a half-filled d-subshell.<br><br>**or**<br><br>It has a half-filled d-orbital. | 1 | A mark should not be awarded for "It has an unpaired electron". |
| | (b) | (i) | 4/four | 1 | Zero marks are awarded for +4/4+/−4/4−/IV. |

| Question | | | Expected response | Max mark | Additional guidance |
|---|---|---|---|---|---|
| | | (ii) | Flame test<br><br>**or**<br><br>Atomic absorption<br><br>**or**<br><br>Atomic emission | 1 | |
| | | (iii) | This is an open-ended question.<br><br>**1 mark:** The student has demonstrated, at an appropriate level, a limited understanding of the chemistry involved. The student has made some statement(s) which is/are relevant to the situation, showing that at least a little of the chemistry within the problem is understood.<br><br>**2 marks:** The student has demonstrated, at an appropriate level, a reasonable understanding of the chemistry involved. The student makes some statement(s) which is/are relevant to the situation, showing that the problem is understood.<br><br>**3 marks:** The maximum available mark would be awarded to a student who has demonstrated, at an appropriate level, a good understanding, of the chemistry involved. The student shows a good comprehension of the chemistry of the situation and has provided a logically correct answer to the question posed. This type of response might include a statement of the principles involved, a relationship or an equation, and the application of these to respond to the problem. This does not mean the answer has to be what might be termed an "excellent" answer **or** a "complete" one. | 3 | Zero marks should be awarded if:<br>The student has demonstrated no understanding of the chemistry involved at an appropriate level. There is no evidence that the student has recognised the area of chemistry involved or has given any statement of a relevant chemistry principle. This mark would also be given when the student merely restates the chemistry given in the question. |
| **4.** | **(a)** | **(i)** | The exact mass should be known/measured and should be close to 4·25 g.<br><br>**or**<br><br>The mass should be around 4·25 g but with an accurate reading. | 1 | |

| Question | | | Expected response | Max mark | Additional guidance |
|---|---|---|---|---|---|
| | | (ii) | Dissolve/make a solution of the silver nitrate in distilled/deionised water (in a beaker). Transfer the solution/it and the rinsings (to the standard/volumetric flask). **(1)**<br><br>Make (the solution) up to the mark in a standard/volumetric flask (with distilled/deionised water). **(1)** | 2 | The first mark should not be awarded if the solid is washed directly into the flask.<br><br>Only one mention of distilled/deionised water is required.<br><br>Only one mention of standard/volumetric flask is required. |
| | | (iii) | Titrate a larger sample (of the seawater).<br><br>**or**<br><br>Dilute the standard silver nitrate solution.<br><br>**or**<br><br>Prepare **or** use a lower concentration of silver nitrate solution.<br><br>**or**<br><br>Dilute the seawater less.<br><br>**or**<br><br>Use a micro-burette.<br><br>**or**<br><br>Use class A glassware. | 1 | A general statement such as "use more accurate apparatus" should not be awarded this mark. |
| | (b) | (i) | Vacuum filtration<br><br>**or**<br><br>Acceptable diagram<br><br>**or**<br><br>Filtration under suction<br><br>**or**<br><br>Fluted filter paper | 1 | A mark should not be awarded for "use a Buchner funnel/flask" without further explanation. |

| Question | | | Expected response | Max mark | Additional guidance |
|---|---|---|---|---|---|
| | | (ii) | To check the reaction is complete.<br>**or**<br>To check all chloride ions have reacted.<br>**or**<br>To check that no more precipitate is formed.<br>**or**<br>If there is a precipitate the reaction is not complete. | 1 | A mark should not be awarded for "to see if there is excess reactant" on its own.<br>An answer that refers only to chlorine should be awarded zero marks. |
| | (c) | | Titration can be used with lower chloride concentrations.<br>**or**<br>Gravimetric method would produce too little/no precipitate. | 1 | If an answer refers to chlorine this should be ignored. |
| 5. | (a) | | $K_a = \frac{[C_6H_5CH(OH)COO^-][H_3O^+]}{[C_6H_5CH(OH)COOH]}$ | 1 | Award mark for K without subscript a.<br>$[H_2O]$ should not be included.<br>If state symbols are included they must be correct but do not need brackets.<br>All square brackets and charges must be included. |
| | (b) | (i) | 0·658 mol $l^{-1}$ **(2)**<br>**or**<br>moles of mandelic acid<br>10g/152g = 0·0658 **(1)**<br>concentration of mandelic acid<br>0·0658/0·100 = 0·658 **(1)** | 2 | 0·66/0·6579/0·65789 are also acceptable.<br>Units not needed for final answer but must be correct if given. |
| | | (ii) | 1·97 **(3)**<br>**or**<br>pH = ½p$K_a$ – ½logc **(1)**<br>p$K_a$ (–log$K_a$ = 3·75) **(1)**<br>pH = 1·875 – (–0·0909)<br>= 1·97 **(1)**<br>**or**<br>$[H^+] = \sqrt{(K_a c)}$ **(1)**<br>$[H^+] = \sqrt{(1{\cdot}78 \times 10^{-4} \times 0{\cdot}658)}$<br>= 0·0108 **(1)**<br>pH = 1·97 **(1)** | 3 | 2·0/1·966/1·9659 are also acceptable.<br>The mark is not awarded for a final answer of pH2 (too few significant figures).<br>Allow follow through, including use of $K_a$ instead of p$K_a$ for third mark.<br>If incorrect equation is used, then maximum one mark can be awarded for use of the correct pKa value. |

| Question | | | Expected response | Max mark | Additional guidance |
|---|---|---|---|---|---|
| 6. | (a) | | 196 kJ $mol^{-1}$ (2)<br><br>**or**<br><br>$E = \frac{Lhc}{\lambda}$ (1)<br><br>**or**<br><br>$= \frac{6{\cdot}02 \times 10^{23} \times 6{\cdot}63 \times 10^{-34} \times 3{\cdot}00 \times 10^{8}}{610 \times 10^{-9}}$<br><br>$= 1{\cdot}96 \times 10^{5}$<br><br>= 196 kJ $mol^{-1}$ (1) | 2 | 200/196·3/196·29 are also acceptable.<br><br>Units not needed for final answer but must be correct if given. |
| | (b) | (i) (A) | $2^{nd}$ order/2/$[ClO_2]^2$ | 1 | |
| | | (i) (B) | $1^{st}$ order/1/$[OH^-]^1$ | 1 | Mark not awarded for $[OH^-]$. |
| | | (ii) | Rate = k $[ClO_2]^2$ $[OH^-]$ | 1 | Follow through allowed.<br><br>Mark not awarded for capital K. |
| | | (iii) | 230 $l^2$ $mol^{-2}$ $s^{-1}$ (2)<br><br>**or**<br><br>$k = \frac{2{\cdot}48 \times 10^{-2}}{[6{\cdot}00 \times 10^{-2}]^2 \times [3{\cdot}00 \times 10^{-2}]}$<br><br>= 230 (1)<br><br>**or**<br><br>$l^2$ $mol^{-2}$ $s^{-1}$ (1) | 2 | 200/229.6/229.63 are also acceptable.<br><br>Any order of correct units is acceptable.<br><br>Follow through applies.<br><br>Units and value must be consistent with answer from **(b) (ii)**. |
| 7. | (a) | | Ethanal/the keto form/left hand side/ reactant | 1 | |
| | (b) | (i) | $H_3C$, H, $CH_3$<br>C — C (=O)<br>$H_3C—CH_2$ | 1 | |
| | | (ii) | A racemic mixture is forming.<br><br>**or**<br><br>(When the enol form converts to the keto) the other enantiomer/optical isomer can be formed. | 1 | Mark not awarded if molecules described as tautomers. |
| | | (iii) | O | 1 | Any orientation is accepted. |

| Question | | | Expected response | | Max mark | Additional guidance |
|---|---|---|---|---|---|---|
| | (c) | | 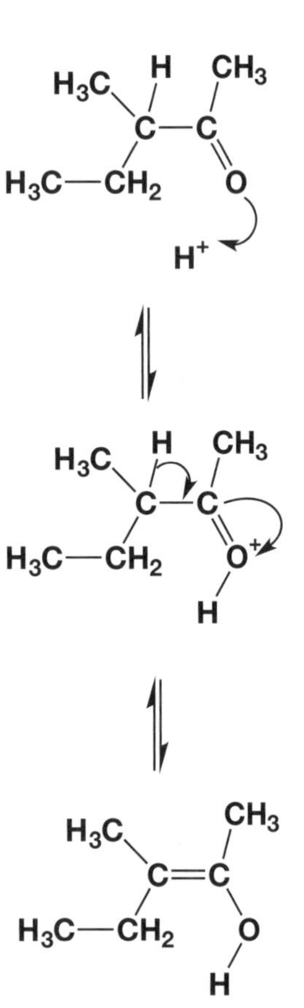 Mechanism following example given in question 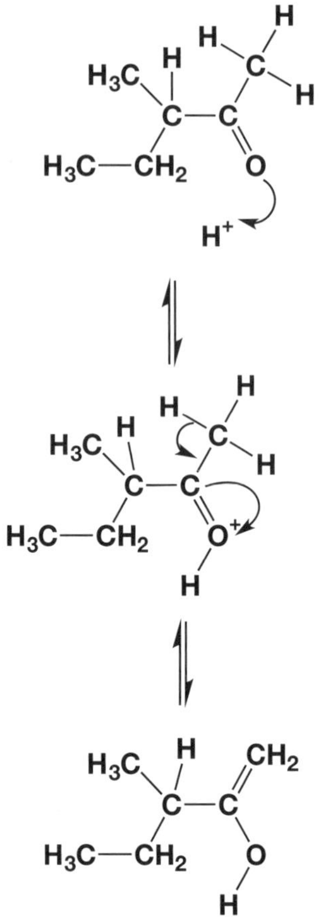 | Correct alternative mechanism | 3 | 1 for product.<br><br>1 for the intermediate – positive charge must be shown.<br><br>1 for the curly arrows – all 3 must be correct and whole headed arrows must be used.<br><br>$C_2H_5$ is acceptable and can be drawn in any position.<br><br>If bond is drawn to wrong part of the alkyl group in the intermediate **or** product, then mark is not awarded. This would only be done once per question. |
| 8. | (a) | | An agonist is a molecule which behaves like/mimics/enhances/triggers the natural response (of the body).<br><br>**or**<br><br>An agonist produces a response similar to the (body's) natural active compound. | | 1 | Do not accept a response which only restates the question stem eg "stimulates receptors". |
| | (b) | | Catalyst | | 1 | Ignore references to homogeneous and heterogeneous. |
| | (c) | | UV (Light) | | 1 | Light on its own is not acceptable. |
| | (d) | | (Nucleophilic) substitution | | 1 | $S_N1$ **or** $S_N2$ would be acceptable.<br><br>Mark is not awarded for electrophilic substitution. |
| | (e) | | $LiAlH_4$ **or** lithium aluminium hydride **or** Lithal<br><br>**or**<br><br>$NaBH_4$ **or** sodium borohydride | | 1 | Mark is not awarded for LAH. |

| Question | | | Expected response | Max mark | Additional guidance |
|---|---|---|---|---|---|
| | (f) | | This is an open-ended question.<br><br>**1 mark:** The student has demonstrated, at an appropriate level, a limited understanding of the chemistry involved. The student has made some statement(s) which is/are relevant to the situation, showing that at least a little of the chemistry within the problem is understood.<br><br>**2 marks:** The student has demonstrated, at an appropriate level, a reasonable understanding of the chemistry involved. The student makes some statement(s) which is/are relevant to the situation, showing that the problem is understood.<br><br>**3 marks:** The maximum available mark would be awarded to a student who has demonstrated, at an appropriate level, a good understanding, of the chemistry involved. The student shows a good comprehension of the chemistry of the situation and has provided a logically correct answer to the question posed. This type of response might include a statement of the principles involved, a relationship **or** an equation, and the application of these to respond to the problem. This does not mean the answer has to be what might be termed an "excellent" answer **or** a "complete" one. | 3 | Zero marks should be awarded if:<br>The student has demonstrated no understanding of the chemistry involved at an appropriate level. There is no evidence that the student has recognised the area of chemistry involved **or** has given any statement of a relevant chemistry principle. This mark would also be given when the student merely restates the chemistry given in the question. |
| 9. | (a) | (i) | $C_9H_{10}O_3$ | 1 | Any order is acceptable. |
| | | (ii) | $sp^2$ | 1 | The "2" must follow "sp". |
| | | (iii) | Orbitals overlap sideways<br><br>**or**<br><br>Orbitals bond side-on<br><br>**or**<br><br>A suitable diagram | 1 | No mark is awarded if any mention **or** drawing of s orbitals.<br>No mark is awarded for an answer that refers to molecular orbitals overlapping. |
| | (b) | (i) | Suitable diagram showing a workable method of condensing the vapour back into the reaction vessel. | 1 | Diagram should<br>• be cross-sectional with inner wall shown.<br>• be an open system.<br>• be sealed around the flask neck.<br>• have water going in at bottom and out at top. |

| Question | | | Expected response | Max mark | Additional guidance |
|---|---|---|---|---|---|
| | | (ii) | The (named) product/products are soluble/ miscible/have dissolved.<br><br>**or**<br><br>There are no reactants left.<br><br>**or**<br><br>There is only product left. | 1 | A mark should not be awarded for<br>"The products are miscible with/soluble in each other" without further explanation |
| | | (iii) | The (4-hydroxybenzoate) ion from the salt removes/reacts with $H^+$ from the water.<br><br>**or**<br><br>Conjugate base of a weak acid, removes/ reacts with $H^+$ from the water. **(1)**<br><br>This results in the water equilibrium shifting to the right hand side.<br><br>**or**<br><br>Shifting to the left hand side if candidate has written an equilibrium reaction with ions on the left hand side.<br><br>**or**<br><br>This results in the water equilibrium producing an excess of $OH^-$ ions. **(1)** | 2 | A mark should not be awarded for "It is the salt of a strong base and a weak acid" without further explanation. |
| | | (iv) | Any two from:<br><br>doesn't react with solute/reactivity<br><br>**or**<br><br>being more soluble in the hot solvent than in the cold<br><br>**or**<br><br>impurities to be soluble/insoluble in both hot and cold solvents/solubility of the impurities in it<br><br>**or**<br><br>boiling point<br><br>**or**<br><br>polarity. | 2 | Marks are not awarded for remove impurities on it is own.<br><br>A mark should not be awarded for choosing a solvent with low boiling point. |

| Question | | | Expected response | Max mark | Additional guidance |
|---|---|---|---|---|---|
| | | (v) | 3·85 g (2)<br><br>**or**<br><br>77·5% = 2·48<br><br>$100\% = \frac{2 \cdot 48}{0 \cdot 775} = 3 \cdot 20$ (1)<br><br>**then:**<br>1 mole 4-hydroxybenzoic acid = 138 g<br><br>$3 \cdot 20\ g = \frac{3 \cdot 20}{138} = 0 \cdot 0232$ moles<br><br>1 mole 4-hydroxybenzoic acid is produced from 1 mole ethylparaben.<br><br>0·0232 moles ethylparaben required.<br><br>1 mole ethylparaben = 166 g<br><br>0·0232 moles = 0·0232 × 166<br><br>= 3·85 g (1)<br><br>**or then:**<br><br>4-hydroxybenzoic acid : ethylparaben<br><br>1 mole : 1 mole<br><br>138 g : 166 g<br><br>$3 \cdot 20\,g : \frac{3 \cdot 20 \times 166}{138}$<br><br>= 3·85 g (1)<br><br>**or**<br><br>2·48/138 = 0·01797<br>0·01797 × 166<br>= 2·9832 (1)<br><br>2·9832/0·775<br>= 3·85 g (1) | 2 | 3·9 g/3·849 g/3·8493 g are also acceptable.<br><br>Correct unit, g **or** grams, is required for the second mark. |

| Question | | | Expected response | Max mark | Additional guidance |
|---|---|---|---|---|---|
| **10.** | **(a)** | | $C_3H_4O_2$ **(2)**<br>**or**<br>C 50·0/12 : 4·167<br>H 5·60/1 : 5·60<br>O 44·4/16 : 2·78 **(1)**<br>$C_3H_4O_2$ **(1)** | 2 | Any order is acceptable.<br>1 mark for correct numbers of moles.<br>1 mark for a correct formula from calculated number of moles. |
| | **(b)** | | C = O (stretch)<br>**or**<br>Carbonyl<br>**or**<br>C=O (stretch) carboxyl/carboxylic acid | 1 | A mark should not be awarded for carboxyl **or** carboxylic acid <u>by itself</u>. |
| | **(c)** | **(i)** | $C_3H_4O_2$ | 1 | Any order is acceptable. |
| | | **(ii)** | $[C_2H_3]^+$<br>**or**<br>$^+C_2H_3$<br>**or**<br>$[CH_2CH]^+$<br>**or**<br>Correct full structural formula of above | 1 | The positive charge must be shown and not placed on a hydrogen atom eg not $C_2H_3^+$.<br>A mark is not awarded for $[CH_3C]^+$. |
| | **(d)** | | H–C(H)=C(H)–C(=O)OH | 1 | Correct full structural, shortened structural **or** skeletal formulae can all be accepted.<br>Allow correct follow though from ALL evidence that the candidate has written for parts **(a)**, **(b)** and **(c)** of the question. |

# Acknowledgements

Hodder Gibson would like to thank SQA for use of any past exam questions that may have been used in model papers, whether amended or in original form.